Becoming a Successful Scientist

Strategic Thinking for Scientific Discovery

Scientific research requires both innovation and attention to detail, clever breakthroughs and routine procedures. This indispensable guide gives students and researchers across all scientific disciplines practical advice on how to succeed. All types of scientific careers are discussed, from those in industry and academia to consulting, with emphasis on how scientists spend their time and the skills that are needed to be productive. Strategic thinking, creativity, and problem solving, the central keys to success in research, are all explored. The reader is shown how to enhance the creative process in science, how one goes about making discoveries, putting together the solution to a complex problem and then testing the solution obtained. The social dimension of science is also discussed from the development and execution of a scientific research program to publishing papers, as well as issues of ethics and science policy.

CRAIG LOEHLE is Principal Scientist at the National Council for Air and Stream Improvement, Inc. in Illinois. His published work has spanned topics including forestry, ecology, conservation, statistics, simulation, climate change, and optimization. He is the author of *Thinking Strategically* (Cambridge, 1996).

Becoming a Successful Scientist

Strategic Thinking for Scientific Discovery

CRAIG LOEHLE
National Council for Air and Stream Improvement Inc., Naperville, IL

CAMBRIDGE UNIVERSITY PRESS
Cambridge, New York, Melbourne, Madrid, Cape Town, Singapore,
São Paulo, Delhi, Dubai, Tokyo

Cambridge University Press
The Edinburgh Building, Cambridge CB2 8RU, UK

Published in the United States of America by
Cambridge University Press, New York

www.cambridge.org
Information on this title: www.cambridge.org/9780521513616

First published 2010

Printed in the United Kingdom at the University Press, Cambridge

A catalog record for this publication is available from the British Library

ISBN 978-0-521-51361-6 Hardback
ISBN 978-0-521-73506-3 Paperback

Dedicated to my dear wife Neda and to my children Sholeh, Gloria, and Niki

Contents

Acknowledgments

Thanks to Tim Benton and several anonymous reviewers for helpful comments on the entire book. Drafts of various chapters were reviewed by Egolfs Bakuzis, Henry H. Bauer, Kent Cavender-Bares, Dan Herms, Rolfe Leary, D. Bengston, John Cairns, I. J. Good, Robert Lackey, Robin O'Malley, D. Orth, M. C. Rauscher, James Schear, David Schlesinger, D. K. Simonton, Robert J. Sternberg, Jacob Weiner, and R. Michael Miller. Their assistance is greatly appreciated. Various chapters and portions of this book are revised from the following sources and used by permission: Loehle, C. 1990. "A guide to increased creativity in research." *BioScience* **40**:123–129. © American Institute of Biological Sciences. Loehle, C. 1994. "A critical path analysis." *The Journal of Creative Behavior* **28**:33. © The Creative Education Foundation. Loehle, C. 1994. "Discovery as a process." *The Journal of Creative Behavior* **28**:239. © The Creative Education Foundation. Loehle, C. 1994. "Creativity the divine gift" from *On the Shoulders of Giants*, George Ronald, Oxford.

1

A science career

Science is a curious profession. It is relatively easy to get into, but much harder to be truly successful at. There are many different paths to success and just as many ways to fail. Many who have an interest in science in school never find their way into a science career. Many who do get an advanced degree in science are never able to get a grant or conduct a successful research program and may leave the field after a while. Some are tempted to cut corners and thus ruin their careers. Even those who have a science job may not be secure in their abilities or their productivity.

This book has four parts. In Chapter 1, science as a career is explored. What do scientists in different fields study? What skills are needed? How do scientists spend their time? How do you choose the right career path? Chapters 2 and 3 cover the ins and outs of creativity and problem solving, the central keys to success in research. Chapter 4 discusses the social dimension of being a scientist.

The most difficult part of a scientist's job is conducting research. Huge amounts of time are wasted in science experiments that fail, ideas that don't pan out, and papers that are never finished. Effort is wasted on proposals that don't get funded and projects whose results are never published. Even published studies are often flawed. Why?

Scientists study science, not psychology, but many of the tricks (and pitfalls) of conducting research are mental. In textbooks a few classic experiments are described (usually only the successful ones) and the student does an apprenticeship (graduate school) with a working scientist. This is not enough. In military training, just learning how to shoot a gun is not enough for an officer. They spend a huge amount of time learning about strategy, tactics, and logistics. But scientists do not. They study calculus and physics and genetics. Yet, for the scientist the enemy is much more subtle: disorganization, mental bias, failure of imagination, fear of disapproval, poor time management, etc. These ideas are explored in Chapters 2 and 3 of the book.

In addition to scientists at all levels, others can benefit from this book. Engineers share much with basic scientists and virtually everything in this book applies to their work as well. Those who interact with scientists as employers or in other roles may gain insight into the field. Psychologists may benefit from reading about creativity and cognition as seen by a practitioner of science rather than an artist or musician. Finally, educators can learn about creativity and problem solving from an applied field with an eye toward enhancing science education.

In certain types of problem solving, special skills are learned which are then applied repeatedly. For example, in accounting one learns certain procedures for keeping books and doing computations, but the component skills are specific to a well-defined domain: accounting. In contrast, scientific thinking involves the integration of several types of mental skills and techniques, as well as certain habits and attitudes, in the context of defining the problem to be solved from an initially ambiguous sea of unconnected data, and then solving it. There is an element of risk in scientific problem solving because complexity causes uncertainty. The acquisition of the skills presented in this book, and their integration, will help reduce risk and increase problem solving success.

While there have been many books about creativity and problem solving, they are mostly about problem solving minutiae, such as the use of analogy, visualization, generation of novelty, brainstorming, lateral thinking, and free association. We may say that these component skills are like the ability to saw, the ability to hammer a nail, and the ability to use a drill, without any skill in reading blueprints or an understanding of how an entire house fits together. While a collection of low-level skills will enable you to build a bird house, they do not allow you to build an office building. To make another comparison, brainstorming may help you come up with a name for a new product or an ad campaign slogan, but it will not help you compose a symphony or build a space shuttle. This book goes beyond brainstorming and describes the tools needed for both generating new ideas and for carrying them through to a completed product.

Gardner (1983) proposed that there are discrete dimensions of intelligence, such as linguistic, musical, mathematical, and spatial mental abilities that are relatively independent of one another and that are not necessarily measured by a general intelligence quotient (IQ). He points out that IQ mainly measures linguistic and logical/mathematical abilities. Musical aptitude is clearly not tested by IQ tests. I believe there is also a dimension of strategic intelligence. This dimension of intelligence comprises a flair for planning ahead and finding the best route or scenario to obtain an advantageous outcome. A person high on this dimension is good at planning a trip and does not often forget to pack something in his luggage. Such people are likely good at board games and poker (though such games may not motivate them because the outcome seems too trivial), and are also likely to be good at making career moves. The person low on this dimension goes to the laundry without soap, gets in the shower without a towel, and goes to the store without a list. Such a person finds a job but has no concept of a career path. It is really best not to go camping with such a person, because they will end up sleeping in your tent, since they didn't bring any tent pegs. They are usually without a clue and are always getting surprised by outcomes that do not surprise others. Such people go to pieces when faced with logistics problems such as organizing their desk, packing for a trip, or reordering the garage. I believe this dimension is independent of other dimensions of intelligence because I have known otherwise intelligent people who are absolutely incapable of planning ahead or anticipating the consequences of actions. This aspect of intelligence is not very amenable to pencil and paper diagnostic testing, which is why I believe it has not been identified and studied previously. This book explores the strategic dimension of mental reasoning and problem solving in the context of scientific research, where it is a particularly critical skill. Specifically involved in this are the identification of the mind's strengths and weaknesses, understanding how cognitive processes operate, and learning how one can obtain reliable information and solve complex problems, how new ideas are generated and tested, and how real world complexity may be dealt with. This book explores these issues and provides training for the strategic dimension of intellectual reasoning, a key dimension for success but one largely overlooked by our educational system. This book is not an academic treatise, but rather is a guide to applying strategic thinking skills in the context of conducting research.

The importance of strategic thinking can be demonstrated as follows. When the frontal lobes are damaged or removed, the IQ of the person remains unaffected and they may even remain at the genius level if they were at this level before the removal. However, the person loses all initiative and the ability to solve novel problems. They will score as well as before in IQ tests, do crossword puzzles and math problems, etc., but will not seek out and solve new problems, such as deciding to remodel the kitchen or invent something. This is exactly the set of symptoms that describes the mindless government official or the corporate drone: they have a college degree and appear smart but they are unable to take initiative and withdraw in fear from novelty. These people have not had an actual lobotomy, but they have been trained and rewarded in such a way that initiative has been squashed. It is not hard to create a drone: merely scoff at all new ideas, have complex procedures that must be followed to the letter, punish mistakes severely, reward conformance, and require approval for every action. The drone can solve simple problems such as accounting problems, arranging meetings, writing a descriptive report, and doing defined technical tasks, as long as the work is defined for them, but they can not create novelty, overcome outdated methods of operating, identify problems with existing systems, or create new concepts or products. For such tasks strategic thinking is required. Technical proficiency and the possession of a college degree is no more a guarantee of strategic thinking than is the IQ score of the lobotomy patient an indicator of their ability to function.

We may further note that the types of problems used in both IQ tests and in most creativity training are contrived and mostly involve the linguistic and logical dimensions of intelligence. The problems typically involve short linguistic riddles (x is to y as z is to what), comparisons, exclusions, analogies, etc., and simple logical operations (short computations). Training in problem solving usually involves simple puzzles such as word problems (a train leaves city P at 10:00 a.m. and . . .). It has been shown, however, that while such tests predict success in school, there is no correlation with success later in life (Gardner, 1983). That is, one can do quite well on standardized tests and get good grades in school, but be incapable of innovation or of dealing with complexity. This is because real life and the production of goods of value requires strategic thinking and creativity, neither of which is either tested by IQ tests or fully developed by current schooling practices, nor is real creativity equivalent to the pure generation of novel responses.

The potential benefit from the application of the information presented in this book is enormous. Gilbert (1978), for example, has documented the huge range of observed productivities among workers. Whether we are looking at academic productivity (publications), computer programming (lines of correct code), artistic output, sales, or any other endeavor, the most productive individual within a job category is often at least 10 times as productive as the average worker, and sometimes as much as 30 times. Whereas in the realm of creative output some hold quality up in contrast to quantity, there is actually more often a correlation between the two: the most innovative individuals often produce the most (Simonton, 1988). This is because the same skills that enable truly innovative work to be done also enhance productivity. The success of the individual professional certainly depends on the frequent production of innovative work. Whereas most artists produce a few to a dozen paintings a year, Andy Warhol and Picasso filled warehouses with their work. While most academics write one or two papers per year (or less), some write a book each year. The same applies to inventors, architects, software designers, or any other profession. It applies particularly to research. This difference in productivity is equivalent to that produced by the industrial revolution or the introduction of computers. Might it not be that such high levels of output could be more generally achievable with the right training? In sports, coaching and training regimes have become a science, with the consequence that the range of performance is usually close to 2 or even less (for example, in professional baseball the record for home runs is only twice that of the average major league player). The range of performance on other tasks, being so wide, means that huge improvements in productivity are possible among those who are less productive. The time spent fixing prior mistakes, spinning one's wheels, doing tasks inefficiently, and doing the wrong task add up to an easy potential doubling of productivity for almost anyone doing any type of nonroutine intellectual work. When the quality of the finished product is considered, there is room for further improvement, making an overall increase in value of an order of magnitude within reach for any scientist.

Why do we think that basketball players or tennis players need a coach but no one else does? That intensive training in technique can lead to a top gymnastic performance we do not doubt, but it never seems to occur to us that a top scientific performance can similarly benefit from coaching. And yet, sadly, today one can not count on one's corporation or university to provide such training and productivity enhancement. In the name of keeping costs down, companies have rejected the idea that individuals should be groomed for rapid advancement by providing guidance, feedback, and special work experiences and training. Instead, the idea has become popular that large numbers of employees should be just kept in the job for which they were hired. Many colleges today, for example, use part-time outside instructors for as much as one third of their classes. Such individuals do not receive any career guidance, have no time devoted to research or professional development, do not have laboratory space, can not have graduate students, can not get promoted, and are not funded to attend conferences; that is, they are in a dead end job. The same is true in many corporations today, where opportunities for advancement have become few and far between. This style of management makes perfect sense from the short-term bottom line perspective, but the rejection of the concept of training people for advancement is that the entire process of increasing personnel productivity is becoming neglected. In my career, over many years as a software developer and research scientist with six different organizations, not one of my supervisors ever came into my office to ask how the work was going or offered a single tip on how to be more productive, creative, or effective. My sole feedback was at the annual review where I was always told that I was doing just fine and to keep it up. Even if completely true, and not just a cop-out from a boss who wants to avoid the performance appraisal process, such feedback is not very helpful for doing better in the future. Many supervisors actually have a disincentive to providing good career advice: if they increase the productivity of their employees, then the employees will expect a raise or even a promotion, which they have been told will not be provided. In this climate, if the employee (i.e. the reader of this book) is to get ahead, he must become noticeably more proficient, talented, productive, and creative so as to stand out from the crowd. To do this, he needs a coach. This book can be your personal coach in creativity, problem solving technique, work habits, and productivity.

There are three pillars of scientific productivity: skill, motivation, and strategic use of time and effort. Skill is what one acquires in school and what one polishes with practice. This includes facts, manual skills such as mixing of paints for a painter and soldering for an electrician, and technological mastery of such tools as spreadsheets and databases. Skill alone only guarantees one a job doing work for someone else as a cog in the machine of a large organization, but does not guarantee high quality work or outstanding performance. Motivation is a key component of success that is generally not taught in school. Few have succeeded without significant motivation, because success requires sustained and substantial effort. Many motivational books by successful business leaders have been written, and these books can be very helpful for increasing motivation, perseverance, and effort. However, motivation alone is not enough because many unimaginative people put in very long hours doing pointless tasks and producing little of value, and companies full of executives putting in 12-hour days have nevertheless gone bankrupt. The third pillar, strategic use of time and effort, is the ultimate key to success, though it depends on the first two being firmly in place. The strategic dimension is what allows one to choose the right problem to solve, to solve it in a cost-effective way, to use resources efficiently, and to be innovative and productive. The strategic dimension, not merely effort, is what accounts for the huge productivity differences noted by Gilbert (1978). In the absence of a concept of strategic use of time, many organizations reward effort rather than output, with the result being that people put in long hours to impress the boss, while being very ineffective in their use of time and perhaps without even producing anything tangible. The combination of skill, motivation, and strategic use of effort can lead to astonishing levels of productivity. This book is based on these three pillars and aims to point the way to such levels of productivity.

One may ask whether strategic thinking alone is sufficient. Of course not. In a world full of downsizing and changing technologies, there is no guarantee of success or of permanent employment. Nor is one always in a position that one's ideas can be carried out. Just because you invent a new product doesn't mean that your company will develop it. However, it is especially under these conditions of uncertainty, where every professional has become a consultant, that optimal output of innovative, high quality work has become most imperative. A number of other issues impact the success of a scientist, including corporate culture, relations with bosses and subordinates, concepts of teaming, and project management. While many of these issues are touched on in this book, the focus is on the performance of the individual: what can you, as an individual, do to become more effective, more innovative, and more productive.

It is useful to contrast this book with Peters and Waterman's *Excellence* (1982), which identified organizational structures and management strategies that have been proven by the test of time (i.e. these companies make money). If you are fortunate enough to work for one of these high performing companies, you will find that the work style promoted in this book will likely be encouraged and the enhanced productivity engendered by these techniques will be rewarded. If you are working for a loser, a company with a bad attitude and a cramped style, then you are probably facing downsizing and need to hone your strategic faculties to get out before you are laid off. If you are a consultant or entrepreneur, then you need this book for your very survival. While Peters and Waterman's book helps one to understand the behavior of the company one works for, most professionals are not in a high enough position to alter the corporate culture. However, one is in complete control of one's own performance. In whatever setting, it is personally better to be creative and productive, even if in the short term it does not seem as if this will be rewarded.

1.1 WHY SCIENCE?

Nature is a gigantic puzzle, and scientists have the unique opportunity to try to put the pieces together. This process of puzzle solving can be both aggravating and rewarding. It is infinitely interesting and engaging. There is scope for exercising creativity and for self-directed work. While demanding, the work is ultimately rewarding. For me at least, once I have started working on a project, I think about it all the time. Each subproblem that gets solved is both exciting and satisfying.

Making a career in science is not necessarily straightforward, however. The media depictions of scientists are very limited and do not illustrate most of the careers available to scientists, nor do they accurately show what they do all day. There are many possible fields, specialties, and career paths, not just the well-known job of university professor. This chapter introduces the types of tasks that engage scientists, the career paths available to them, and the skills and aptitudes that a scientist needs to have or to develop.

What scientists do

In the popular imagination, scientists make discoveries. While this is part of it, there are actually many activities that scientists engage in that are not "Eureka" moments.

There is often a separation between those who propose a theory and those who test it. This is party due to temperament. The dreamer makes a better theoretician than experimentalist. In addition, once a theory is proposed it can take lots of experiments to test it properly. Who has made the discovery: the one who proposed the theory, or the one(s) who tested it?

A major part of the process of science is the development of methods, tools, and instruments. We can think of remote sensing technology, the electron microscope, and growth chambers as tools that enabled new knowledge to be gathered. The development and testing of such new tools is "science," though it is not a discovery. An additional type of tool is the mathematical or statistical tool. For example, the randomized block experimental design with its accompanying statistical tests is an essential tool in some fields.

The gathering and cataloging of basic data are likewise part of the scientific enterprise. For example, protein databases, catalogs of species, geologic maps, historical climate data, and gene sequences are all useful to the scientific enterprise even though those who collect and maintain this data are not making "discoveries."

The cases just presented represent the infrastructure of science, the machines, lab techniques, mathematics, and databases that enable discovery. Many scientists spend part or all of their time working on infrastructure, and this work is in fact "science."

Anyone conducting scientific studies must eventually communicate their results in order for them to become part of the common body of knowledge. Thus communication of results is critical. Those who love to do research but hate to write or talk about it will usually fail. There are three principal outlets for communicating scientific results: publications, conferences, and seminars. One only needs a plausible abstract to give a talk at a scientific conference, so talks often represent preliminary or tentative results. Some of this work will never make it into print, and if it is not in print, it does not become part of the body of common knowledge. It is thus best not to fool oneself that a conference talk is an adequate outlet for one's results. Seminars have the advantage that there is more time for questions and discussion, but again, are an impermanent outlet for one's work. Thus scientific publications are a critical avenue for communication of scientific research results.

Sometimes scientists are engaged in practical research. They may be asked not to develop general hydrologic theories, but rather to clarify the hydrology of a particular watershed used for a city's water supply. They may conduct a field survey for an endangered species or do a toxicity test for an industrial chemical. All of these are scientific projects, even though they may add little to basic knowledge. Remember, though, that Pasteur's early work was funded by brewers who wanted to know how to prevent their vats from spoiling.

Some scientific work involves synthesis. Usually in written form, a synthesis draws together a body of knowledge, discusses competing views and conflicting studies, and organizes the subject matter. The classic synthesis is the textbook, but review articles and commentaries also fall into this category. The synthesis of a body of knowledge really is a creative process, though it is sometimes disparaged as "just a literature review."

Large science projects require management. For example, large physics experiments such as colliders or neutrino detectors require huge teams which must be managed. These projects must be managed by real scientists, not by professional managers. It is sometimes believed that an alternative career path for a scientist is to go into management, but this can quickly become an excuse to stop doing science and is no longer a science job.

One might think it is obvious that scientists teach, but generally only academic scientists do so. While the university used to be where most scientists spent their career, this is no longer the case. For those who teach, it can consume all their time or be a nice break from the lab and a chance to share their knowledge. Teaching can also be an excuse for not conducting research. Even within a university, many scientists work in research institutes funded by the government, by grants, or by foundations.

Scientists work in groups and they work alone. Ultimately, a scientist must face his experiments and data alone. The average scientist is not a people person. But in the process there are collaborations, both big and small. Often there is a need to combine specialties in order to solve a problem. For example, some wildlife biologists had done field surveys and written up that work. My colleagues and I had some ideas for pulling together landscape data and using it to analyze the field surveys. So, we pulled together all the field people into a loose team, we did the analyses, and everyone was happy (and everyone was a coauthor). I have coauthors that I have never even met. Sometimes very large teams are involved. I was coauthor on a paper with 19 others but this was really quite awkward, because we were all authors rather than having specific subtasks. Some scientists do not work well with others, or always must be in charge. So the way a scientist interacts with colleagues can vary considerably.

As a final note on the types of work that scientists do, the extent to which their work is adventurous varies radically from person to person. An anthropologist may dig up ruins (tedious but mixed with adventure) or study gangs (danger *and* adventure!). On the other hand, he may not go into the field at all. A scientist may work with satellite data (can't visit his instrument), or develop string theory. The extent to which he works with field data, laboratory data, or pure theory will strongly influence his day-to-day work experience.

Careers

In the past, most scientists worked for universities as professors. The Cold War brought big science in the form of the national laboratories, which employed thousands of physicists, chemists, and mathematicians. In addition, public agencies such as the United States Forest Service, National Oceanic and Atmospheric Administration (NOAA), National Aeronautics and Space Administration (NASA), Department of Agriculture, and Centers for Disease Control became havens for basic research. Now, therefore, there are all sorts of career paths for scientists, and many of them never come close to a university or government lab.

There are many ways to think about science careers. One can think about them in a disciplinary fashion, in terms of pure versus applied, or in terms of industry. For example, in a disciplinary sense one might have an interest in genetics. In a pure science context, one could work at a university. But genetics could take one into agriculture (crop improvement), biomedical work (tracking genetic diseases), pharmaceutical work (the genetic link to drug development), or entrepreneurship (developing lab tests for genetic conditions). Thus for a career in science one really needs to think differently about it than just in terms of the academic disciplines. It is useful to review, for several fields, what the possible careers are. This is done next.

Applied science fields such as forestry, agriculture, mining geology, or veterinary are self-evidently related to a profession. They also, however, can lead into unexpected careers. In forestry, there are academic jobs, as one would expect. In addition, there are research-related jobs at all levels of education in government agencies such as the U.S. Forest Service. These jobs involve field surveys, lab and field experiments, remote sensing data analysis, running a greenhouse, etc. There are also hands-on management jobs running tree planting, timber harvesting, and other operations. There are also jobs in conservation and environmental advocacy, which may still involve a research component (or may not). Similarly, agriculture and veterinary (or animal science) degrees can lead into research, industry (e.g. large-scale farming), manufacturing (e.g. pet food, farm equipment), or sales. Jobs in international development are possible.

In the basic biological sciences it can be a little trickier to find a career because they do not feed into a coherent business. Zoology, for example, involves the study of subjects that may not be directly marketable. Zoology that grades into wildlife conservation could lead to wildlife management or conservation jobs. Zoo and museum jobs are few and low paying. Thus zoology as a field resides largely within the university environment.

Botany is very similar to zoology in this regard. Some jobs exist for those who link their botanical interests to agriculture (crop science, weed control, trees and forests). There is even less money available for the study of rare plants than for endangered animals.

The microscale studies of genetics, microbiology, and cell biology used to be similar to zoology and botany in being pure sciences that were largely confined to the university. Over recent decades, however, they have become linked to biomedical research institutes and corporations that produce drugs, diagnostic test equipment, and therapies.

Geology has obvious connections to oil and mineral exploration. This can lead to all types of careers, including mapping, site characterization, and even basic and applied research. Geologists are also involved in hazards identification, such as monitoring volcanoes, tracking earthquake activity, and evaluating landslide risks.

Chemistry has direct linkages to many industries, from chemical manufacturing to pharmaceuticals. The Environmental Protection Agency (EPA) hires chemists. Environmental testing and cleanup firms employ them. Chemists even help produce snack foods, flavors, and makeup (though it may sound more as if alchemists are involved). If aiming at the more applied end of the spectrum, one might want to study chemical engineering.

In all of these examples, there is a spectrum of opportunities ranging from basic science to applied science to industrial. The latter jobs are based on science and can be quite high-tech, but may not involve research per se. Thus one must be clear about whether one really wants to do research or not, and at what level. A field technician who surveys for endangered species is involved in research but is not in charge of it.

In addition to considering the field of study and basic versus applied options, it is helpful to consider possible career paths. For a university professor, the path seems obvious: promotion through the academic ranks. An additional option is to move into administrative positions. This can be done while maintaining an active research program, but often is not. Academics can also branch out into inventing, book publishing, consulting, software development, and entrepreneurship. These secondary endeavors can be a hobby, a distraction, a good income supplement, or a new career, depending on luck and aptitude. Many academics, however, seem blissfully unaware that they could pursue any activities outside of grants and writing papers.

In industrial settings, there are many pathways to upward mobility. One can work one's way up the ladder in the Research & Development department, in sales, in manufacturing, or in other areas. Each of these areas will have different career paths and different opportunities. Finally, entrepreneurship is an option in many fields. Examples of small scientist-led start-up firms with which I am familiar include a food flavors company, a company that developed and runs aquatic water quality software for the U.S. EPA and private firms, statistical and scientific software firms, consulting firms, groundwater testing firms, and many others.

So, how does one decide on a career in science? There are many ways to arrive at a decision, some of which are quite irrational. For example, some individuals find themselves fascinated with ants or birds or fossils from an early age. Why? I have never seen an explanation for this early attraction to a scientific subject. For me it was trees. I can remember the trees in the neighborhood where I lived at age 7 so well that I can tell you what species they were and how tall. Such a fascination with a topic is a natural lead-in to a science career (yes, I ended up studying forestry!), though not all with such a passion find their way into a science career.

A different type of passion is a passion to be outdoors. Exposure to hunting and fishing are often the spurs to this type of passion. Scientific jobs that are also outdoors include geological fieldwork (either prospecting or research), wildlife biology and conservation, and forestry. One should realize, however, that typical jobs that involve fieldwork such as a university position include other duties such as teaching. Fieldwork may end up being possible for only a month or two each year. This can be frustrating if not taken into account when planning a career.

It is often the case that someone is drawn to science because they do well in science classes. This is a reasonable perspective but is not an entirely adequate basis for career choice. One needs to consider what a person in such a field does, not just the class work encountered as a student. Otherwise, one can end up like the person who became an English teacher because they loved books, but then discovered that they hated kids. Posing and solving original problems in pure mathematics is an altogether different thing than solving calculus problems in a class. As another example, a physicist may need to be mechanically inclined and not just able to learn textbook physics. On the other hand, if someone is fascinated by nature but can not do well (due to insufficient aptitude or lack of inclination to study hard) in science classes, then a hobby such as bird watching or fishing is more indicated than a career in science.

A consideration when evaluating a science career is how one feels about working with people. Most university positions involve teaching and advising. If you just want to get lost in the intricacies of string theory, teaching might drive you crazy. Physics experiments and drug development can involve large teams and are not for loners. Because many scientists are attracted to science because it *is* a solitary pursuit, and because one is judged on the work and not on fashion sense or family background, it can be disconcerting to end up working with people in the end. Thus it is helpful to think this issue through and not end up surprised by it.

There is a tremendous range of salaries in science careers. Salaries at small colleges, for example, are similar to those of public school teachers. If one is fixated on a particular organism or topic that is not in wide demand or not connected to some industry, then even obtaining a job can be difficult. Generally speaking, areas of science that are more mathematical and/or that have an associated engineering degree or industry affiliation will pay more. Many of these engineering specialties are essentially sciences with an applied aspect (e.g. chemical engineering, materials science). In these sciences there are opportunities for patents, for consulting, and for entrepreneurship, as well.

There are psychological factors that influence what science career would be the best or most satisfying choice. This is because the demands of the different fields/jobs can be vastly different. For example, to be successful in pure mathematics or theoretical physics requires very high intelligence and mathematical aptitude. Without this level of skill, one can achieve nothing. In other fields of science, creativity and/or hard work will enable the merely intelligent practitioner to be successful.

Degree of focus and persistence are critically related to the type of work demanded in different fields. In order to successfully excavate an archaeological site, one must be able to organize sustained (10+ years) effort on that site. In cancer research or chemistry, some ideas can be tested in hours or weeks and a rapid succession of ideas is needed.

Creativity is a key skill for scientists but the degree to which it is critical varies by field and by job within the field. Those who compile and organize a star database need to be methodical and organized rather than creative. A science teacher needs to be creative in conveying information but less so in discovering it. The role of creativity in research is the topic of the second chapter of this book.

How one deals with pressure is relevant to career choice. Some fields move very fast, with high prestige and grants going to those who are first with a discovery. If someone likes to dream and ponder on a topic, such high pressure fields may not be a good fit.

Many people stumble into a career in science. Where they end up may not be the best fit to either their talents or their psychological needs and strengths. This can lead to restlessness and suboptimal performance. Such adverse outcomes need not happen, because much is known about the requirements of different science careers.

1.2   THE LIFE OF A SCIENTIST

The typical picture of a working scientist is dominated either by the mad scientist in movies or by experiences with science teachers/professors. While the latter experience is more realistic in one way, the interaction one has with teachers provides no information on the daily life of a scientist. There are many facets to a science career, and many different activities, including serving on editorial boards of journals, conducting research, serving on advisory panels, writing, and advising students. Some activities, such as writing grant proposals, are usually viewed as unpleasant but necessary. Many other activities, such as giving talks and conducting research, can be so satisfying as to be almost addictive.

Many scientists pay the bills partly or largely by teaching. It is possible to teach science courses without being a scientist. Examples would be high school and junior college science teachers. The reason is that one must be actively engaged in generating, organizing, and synthesizing new knowledge to really be a scientist. Communicating textbook knowledge to students is divorced from the pursuit of new knowledge. Unfortunately this often means that presentations of the "scientific method" in classrooms are flawed or incomplete. In my view, the best science teachers are those actively engaged in research. Thus pressures at universities to increase teaching loads in order to save money can have a detrimental effect on teaching quality. The focus of this book, however, is the practice of science, thus teaching is not discussed further.

It is worth noting the distinction between scholarship and science. In the liberal arts, a person can spend years learning a discipline. For example, let us say they learn ancient Greek and translate previously lost works. They are certainly scholars, but this is not science. In science, reliable knowledge about the natural world is discovered and systematized. Sciences vary in the reliability of the knowledge they develop. For example, psychology is a mixture of reliable knowledge, preliminary findings, and personal opinion. It is a mistake to expect the same level of reliability from all fields just because they try to be scientific. Physicists using the laws of motion can predict the path of a satellite sent to Jupiter years in advance, but psychologists can only give probabilities that a particular person will develop schizophrenia. This book is focused on those subjects likely to yield to the scientific method and produce repeatable and reliable results.

Research

Research is the central activity that defines what it means to be a scientist. It is what Nobel Prizes are given for. It is also the activity that scientists themselves find most rewarding.

It is called a scientific "method," which would seem to imply that once you learn the method you can go and do science. There is more to the story, however. Some scientists do conduct research using the method but the result is pedestrian and attracts no notice. Others can never seem to come up with ideas or can't complete projects they start. In contrast, the best known scientists often have so many ideas for research that they are limited only by the number of hours in a day. Many scientists only publish a handful of papers in their careers, but the most prolific can pass 1000. Thus, knowing the "method" is no guarantee of success. The psychological and practical aspects of achieving high rates of success in research are the focus of the second part of this book. At this point, however, it is useful to review what the conduct of a research project looks like.

There are many prerequisites to any research program, such as funding, facilities, technicians or students, collaborators, etc. Here I assume that facilities are available. The other issues are discussed later in this chapter.

Scientists get an idea for a research project in various ways. Perhaps someone has proposed a theory and they think they can do an experiment to test it or conversely some newly published data cries out for an explanation. New laboratory devices provide opportunities for collecting novel data. When the microscope was invented, everything looked at was a new discovery. Perhaps the scientist has conceived a new approach to an old problem. In any case, the experiment or analysis has a definite purpose and does not consist of merely looking at stuff. At this point in a project, the scientist is often very excited and anxious to start, because something new promises to be revealed by the planned project.

Carrying out a project may depend on obtaining funding or traveling to some distant place (if perhaps one wishes to study the rain forest). It may also take a very long time. For example, excavating an archaeological site can take years. In most cases there is a delay of a year or more between project conception to completion. During this time one must stay focused on the tasks and keep the goals in mind. Not everyone is suited to such long delays.

Chronologically, data analysis must come after data collection, but a good scientist designs the experiments in the context of the analyses that are planned for the data. If this is not done, the data may not fully meet the needs of the researcher. For example, if there are too few data, certain statistical techniques will not work. In the worst case, no conclusions can be drawn due to a lack of statistical power.

The process of carrying out a study can be exciting. Finding and excavating a dinosaur bone is certainly an exciting event. Chipping away all the surrounding rock, however, is about as tedious as tedious gets. Some scientists deal with the tedium by delegating tasks to volunteers (at archaeological sites) or to graduate students or technicians. Such an approach can lead to errors, however. For this reason, a scientist may choose to conduct his experiments personally.

A scientific study does not always go smoothly. In a theoretical study, one may have trouble getting all the pieces to fit together or may encounter contradictions. To properly analyze the data one may need to learn a new statistical or analysis technique. Equipment can malfunction. These problems can lead to mental detours, delays, and aggravation. The successful scientist persists in the face of these obstacles but sometimes these barriers are not overcome and the project dies. Thus it is critical that a scientist be able to troubleshoot and deal with the details of data, equipment, and procedures.

The process of overcoming project obstacles progresses like a rollercoaster ride: up and up as the pieces come together and back down as another glitch is encountered. You plot the data and it looks like garbage (down you go) but then discover you read the wrong data columns (back up!). For me, this is much more fun than a real rollercoaster, because when the ride is over I really have something.

Writing

After conducting a study, it is necessary to write up the results for publication. Science is a shared endeavor and unless shared it is not a contribution. Leonardo Da Vinci's books on art made a huge impact but his scientific work, though way ahead of his time, had no impact at all because it was never published.

A scientific paper has a standard structure. In the introduction, the problem is stated. Then one presents some context for the work: past work, what is known, what is up for debate. A methods section presents the methods used. This may be supplemented with online material which provides more details. Next, the results are presented, usually without comment. In the discussion, the significance of the results are discussed and they are compared with past work and perhaps with the predictions of competing theories. The if, ands, and buts of the results (x may not apply under such and such conditions, limits of interpretation, etc.) should be covered.

Writing a scientific paper is not easy. Failure to properly or adequately put the work in context with the existing body of knowledge is sufficient to cause the rejection of even a perfect and novel experiment. Muddy or confusing discussion is a lethal flaw. Using the wrong statistical test can also lead to rejection. Obtaining friendly editorial advice can thus be very helpful for a young scientist. However, do not, under any circumstances, seek writing advice (except proof reading) from an English major. Scientific writing is compact and utilitarian and it will horrify your friend the English teacher. Writing a scientific paper is frustrating and rewarding, but more importantly, is necessary.

Every scientist has research projects that fail. Maybe a hurricane hits your field site or it burns up. Maybe there is a confounding factor that makes results hard to interpret. Sometimes everything seems OK but you can't get it past the reviewers. It is important to minimize these failures. The second chapter of this book particularly focuses on this topic.

Different styles of writing are appropriate for different types of studies. In a highly controlled experiment, the writing is spare and utilitarian. It is necessary to explain clearly what methods were used and what analyses were done. The reader of the paper should not be left wondering about procedures. In a theoretical paper, clarity of exposition becomes more important. Frequently in such papers the assumptions are not stated clearly enough and symbols are vaguely defined. This leaves readers unclear about the meaning of the paper and makes it difficult to judge. Subsequent work may dispute the findings based on a misinterpretation of the results. For example, in models used to study animal behavior such as habitat selection it is implicitly assumed (i.e. never stated) that all individuals are being treated as if they are identical. When this assumption is relaxed, very different results may be obtained. Badly written theoretical work may be ignored and is easily rejected by journals.

Ironically, eloquently written theoretical work can be convincing to people even when wrong. This happens fairly often in semihard sciences such as ecology, evolution, medicine, and geology. It happens very often in anthropology and psychology, to their detriment. In these fields, beautiful theories (the noble savage idea, the free love findings of Margaret Mead in Polynesia, theories that mental illness is just a form of creativity) can persist and have pernicious influences on society and on the progress of the field.

A different type of writing is the review paper. The purpose of the review is to survey what is known about a particular topic. The review will attempt to put studies into context and relate them to theory. It might be that certain studies get contradictory results. The review might attempt to explain these contradictions in terms of various confounding factors or methods used. A review might try to formally compare competing theories. This can involve comparing simplifying assumptions, mathematical formalisms, and predictions of the different theories. The writer might try to determine what type of data or experiment would enable the different theories to be distinguished from one another. In a meta-analysis, the scientist combines data from multiple studies to see if more robust results can be obtained. He or she might ask what percent of existing studies find a particular effect (e.g. side effects from a drug, response of people to insufficient sleep, etc.) and commonalities of those that do and don't.

Many journals do not publish reviews or surveys because they are not "novel." If, however, the review becomes a true synthesis or involves a meta-analysis then this restriction can be overcome. Other journals are open to reviews or even specialize in them. To be successful, a review should do more than simply catalog the literature. It should organize and synthesize the literature in a manner that improves the state of knowledge in the field.

Giving talks

Communicating new knowledge is an essential part of being a scientist. While it would seem that a scientific paper is the ideal medium for communication, the scientific talk is in no danger of extinction. There is something about a presentation that conveys more information than one might expect. This is true even though the scientific talk is not as polished as the published version. Thus becoming a good speaker is important both for career advancement and for getting one's ideas recognized and accepted.

The scientific talk largely skips over the literature review that characterizes the scientific paper in favor of a simple statement of the problem being addressed. Word slides should be a reminder to the speaker of what to say, though many speakers read them verbatim, which can be tedious. The particular advantage of a talk is that the speaker can point to and discuss particular aspects of equations, graphs, or diagrams and lead the audience through to a better understanding.

Talks are most commonly given at scientific conferences. At a large conference one might hear hundreds of talks, which can be a little exhausting. After hearing a talk closely related to one's own work it is professionally beneficial to introduce yourself to the speaker at the break and talk about their (and your) research.

Most science departments have a seminar at which their faculty and visiting scholars can give talks. The advantage of this forum is that the speaker may have up to a full hour to speak and take questions, so a deeper treatment of a topic is possible.

Scientists can be dreadful speakers, which is not helpful to their careers. While one can take lessons in public speaking, scientific talks differ significantly from typical public talks in that much more coherence and precision are required. It can be helpful to observe speakers that are obviously popular (that the room is filled with 200 people instead of 20 is a hint). Since scientists can be a little socially oblivious, they may miss clues that they are boring or confusing. A conscious effort to try to read one's audience can be useful in such cases. A hint that you are boring is that people leave in the middle of your talk. A hint that you are confusing is getting questions from the audience that are irrelevant or that you thought you explained.

Professional service

A successful scientist encounters many opportunities to be of service to his profession and to society. At a minimum, he will be asked to review manuscripts for journals. While one need not review every manuscript one is sent, there is an obligation to review more than you publish. This topic is discussed further in the Ethics section (4.2) later. Prominent scientists may be asked to serve on editorial boards or even to be editor in chief of a journal. Some view this as an honor and as a chance to influence the development of their field. Others may view the work as interfering with their research.

Opportunities exist for service in professional societies. In addition to positions as officers, there are many committees that one can participate in (e.g. for awarding prizes or scholarships, for public outreach, etc.). In contrast to editorial work, this type of service offers more chances for developing relationships.

A common venue for professional service is to serve on a panel. Proposals for research grants and candidates for postdoctoral fellowships are both evaluated by panels of experts. These panels are supposed to act objectively but individuals may nevertheless be more favorably disposed toward their friends or students of their friends. A proposal that is too difficult (too much math, new theory, fancy statistics) is unlikely to gain support on a panel unless a panel member acts as an advocate for it and helps explain it. Panel members should do a little background work before coming to the meeting in such cases, but rarely do. It is well to remember that funding agencies are not bound by panel recommendations, which are just that.

Another type of review is the outside review of a lab, experimental facility, or academic department. This can be for accreditation or simply to provide suggestions for improvement. Such reviews can range from *pro forma* to make-or-break outcomes.

Government science programs from NASA to the Department of Energy to the Forest Service make use of outside review panels to provide feedback and advice for research plans and programs. In some cases, these advisory panels are required by law. Their purposes vary, from helping to coordinate activities across programs to brainstorming on a topic. Often the members of such panels are not clearly informed of their mandate, which can lead to aimless discussions. In some cases, the agencies are required to have advisory panels but are not really interested in their advice. Such *pro forma* panels are a waste of the scientist's time and should be avoided if possible (though it can be hard to tell before attending the first meeting what the situation will be).

Scientists have cause to attend various types of meetings. The most visible type is the scientific conference. At these conferences, one can hear scientific results as much as a year before they come out in print. Sometimes chance encounters can lead to research collaborations. Journal editors find an opportunity to meet with their editorial boards and book publishers meet with their authors. Scientific conferences are also a chance to socialize with colleagues. Such meetings can provide a welcome break from research, but too many meetings can be a distraction.

Scientists must often travel to meet with colleagues. Collaborative work often involves arranging for data exchange, writing proposals as a team, supervising the collection of data in many scattered locations, and so on. For example, several studies might be conducted at a biological field station. Periodically these scientists will get together to see how the pieces of the puzzle (studies on trees and birds and insects) all fit together.

While scientific information needs to be made public to be useful and to be acknowledged, it is also sometimes true that scientific discoveries can lead to intellectual property. The types of intellectual property vary widely across different fields and the awareness of the possibility of financially benefitting from intellectual property likewise varies. In some fields, it is possible to obtain patents, which can be quite remunerative. Examples include any field of engineering, physics, chemistry, and biotechnology. In some cases the scientist may start a company to take advantage of his patents. It is good to remember, however, that most patents do not lead to products or to income.

Books are commonly written by scientists. Specialist books do not sell many copies and thus contribute more to a scientist's reputation than to his wallet. Tenure in some departments can even require publishing a book. Textbooks can bring in substantial revenue, but if not used widely may be trivial sources of income. Books on hot topics can become big sellers, such as Stephen Hawking's or E. O. Wilson's books, even if quite technical. Books on psychology or medicine are the biggest sellers in this arena.

Software is a key tool that can embody intellectual property. Examples include statistical packages, mathematical tools, chemical analysis tools, geological profiling tools, etc. Some of these have made their developers rich, others not. In cases where a market is unlikely to exist, the scientist may choose to make it public domain.

In some cases, expertise can not be captured in a patent or software but is still valuable. In such cases a scientist may choose to start a company or do consulting.

There are many surprising ways in which the expertise possessed by a scientist can be valuable. This can be very important for academics on nine-month appointments who need extra income. Archaeologists who live in places with many artifacts (e.g. Rome) may be called on to do surveys. Those in medicine and psychology may have opportunities to testify in court. Statisticians often provide advice to industry or to other scientists. Advice in the form of consulting for industry is very common in the applied sciences (agriculture, forestry, geology, materials science) as well as in physics and chemistry.

In considering whether to consult, it is helpful to remember that the hours in a day are limited. If the pursuit of consulting interferes with doing research, then tenure may not be gained. On the other hand, consulting work can provide the case studies for economics or management science researchers or lead to publications or even patents in other fields, acting thus as an alternative to government grants.

2

The inner game of science

Science is about mathematics and rigor, but the human mind is sloppy and vague. We are prone to jumping to conclusions, prefer short chains of logic, are easily duped by optical illusions (even when told they are illusions beforehand), and are influenced by group think. This section is about the inner game of science, the mental world where discovery and proof take place. This is the terra incognita that few scientists have thought about, but about which they remain ignorant at their peril. A strategic approach to problem solving is presented and applied to help elucidate the solution to the paradox of how our illogical mind can ever produce reliable, logical results.

Because scientific discovery is inherently about open-ended, complex problems, it is not really possible to apply a cookbook approach ("the scientific method"). Rather, I suggest that a strategic approach to problem solving is a more effective approach. For those who are not born strategists, this involves tuning up one's thinking machinery specifically for this type of problem. Specifically, there are three steps to creating successful novel products. First, one must be capable of generating novel (but useful) ideas, of overcoming routine ways of functioning, and of putting together information in new ways. Without this capacity, one is limited to solving problems defined by others (the definition of a drone). Second, one must be able to use this creative capacity to discover/invent something new. This can be a new style of art, a new computer algorithm, a new statistical method, or a new theory. Third, one must understand how to structure work to bring a new creation to fruition. Without this third step, one is just daydreaming.

In this section I show how one enhances the creative process in science, how one goes about making discoveries, how one puts together the solution to a complex problem, how one tests the solution one obtains, and how one enhances internal mental states for optimal performance. What is most noteworthy is that the conventional wisdom in all of these domains fails in the face of complexity. How one goes about achieving insights on simple problems does not provide a reliable guide to the solution of complex ones. Insights on simple problems can easily be all-of-a-piece aha! experiences, matching the standard preparation–incubation–illumination model, but complex problems require structuring, planning, scheduling, tinkering, and iterative improvements and are almost never solvable in a single flash. The type of creativity measured by typical tests is essentially the generation of novel responses and is unrelated to true creativity, which can only be assessed in ambiguous and complex problem domains. The relationship between intelligence and creativity has traditionally been unclear (i.e. little relationship has been found when a strong one should be expected) but this can be explained when the entire problem solving process is analyzed. A critical step, almost ignored in the creativity literature, is reality testing of a new idea or product. Finally, the attitudinal dimension has rarely been explored, but in fact turns out to be crucial and can explain many great failures of problem solving, such as military disasters and engineering fiascoes. Thus in this chapter, I explore the mental operations and overall processes of strategic problem solving. The third chapter of the book covers strategic problem solving in practice, covering problem solving tools and approaches and discussing general principles such as perspective, information transmission, scale, paradox, and bottlenecks.

2.1   STRATEGIC CREATIVITY

An essential component of strategic thinking is strategic creativity. Because strategic thinking essentially deals with complex problems, uncertainty, and ambiguity, and involves finding solutions that others have overlooked, it is essential that the practitioner be creative in his approach. The standard view of creativity, however, is seriously flawed for application to strategic problem solving. Our typical picture of the creative person is the artist or musician, who may have a wild lifestyle, do drugs, and dress in a bizarre fashion. This represents a confounding between the person with a novel lifestyle or a person with flair and someone who is genuinely innovative. Thomas Edison did not wear tie-dye tee-shirts. Whereas a wild lifestyle may help or not hinder (though this is debatable) an artist's expression of ideas, this is not the case in the realm of research and other technical work. In technical and management professions, one needs to be able to think extremely clearly, to be organized, to plan ahead, and to meet deadlines. Wild, magical, and fuzzy thinking are deadly in this context. This in no way means, however, that creativity must be excluded. Far from it. Those professionals who are technically competent but not creative are limited to doing what everyone else already knows how to do. They can run an organization that is already running smoothly, but will not encourage or even allow innovation. They are the managers and technocrats who very competently run a large organization right off a cliff because they either fail to see change or can not find creative solutions to changing conditions (e.g. the managers and engineers at many computer companies who failed to see the trends away from mainframe computers).

Creativity as it is commonly studied and assessed by psychologists usually involves first order creativity only. The types of measures commonly used include the generation of novel responses, use of analogy, ability to incorporate rich imagery into a story, etc. While such measures of creativity may be directly applicable to the artist, especially a modern artist, because novelty per se is a key component of his success, such first order creative responses are almost irrelevant to strategic creativity. The schizophrenic produces highly divergent responses in drawing or word association, but they are usually not useful. For solving real world strategic problems, one must find not merely novelty, but novelty in the context of constraints, tradeoffs, and uncertainty, and the novel solution must be useful. Because of the complexity of the problems, what is required is not just the generation of novel responses and the habit of creative expression, but rather that one take a creative approach to the thinking process itself. That is, the strategic thinker must be creative in the use of his mental faculties. This is strategic creativity.

Innovative entrepreneurs, academics, scientists, engineers, and inventors are held in high regard, but the means by which they achieve innovation are not spelled out in any manual. Courses on the scientific method do not cover creativity. Master of Business Administration (MBA) programs offer no courses on creativity, innovation, or problem solving techniques. Philosophers of science are more concerned with formal theory structure, proof, logic, and epistemology. Karl Popper (1963), for example, invokes the generation of alternative hypotheses but says nothing about where one is to get them. The type of creativity studied by psychologists is generally more applicable to the arts, where any type of novelty is interesting and need not conform to limitations such as feasibility or cost.

The pressures on scientists today oppose truly creative thinking. Pressures on academics to write grants, teach, and publish leave little time for undirected thinking. Industrial laboratories today are far more directed than in the past, particularly where product development costs are high (e.g. in drug development). Thus one must actively counter these anticreative forces to be a successful innovator.

In this chapter, I describe the characteristics of strategic creativity and provide some guidelines for enhancing it. The capacity to generate new ideas is almost universal, but needs to be enhanced to become a significant problem solving asset. This chapter shows how to identify problems worth working on, how to overcome barriers to the generation of new ideas, how to listen to new ideas when they arise, and how to arrange work schedules so as to enhance creative thought. This provides a basis for generating the ideas crucial to the overall strategic problem solving process. The following chapters build on this material with a discussion of the processes of discovery and invention, which are at the heart of innovation, and of generating finished products. An appreciation of both the dimensions of creativity and the process of discovery is the key to creating results that are both novel and useful. Later sections then present specific tools of thought and general principles of problem solving in complex domains.

Choosing a problem

Perhaps the most important single step in the problem solving process is choosing a question to investigate. In contrast to school settings, in the real world one's problems are rarely handed to one already well defined. What most distinguishes those innovators noted by posterity is not their technical skill, but that they chose interesting problems. There is some guidance that may be given in this regard.

Picking fights

Professional work is supposed to be an objective, dispassionate business. After all, one is dealing with numbers and facts and schedules, with machines and networks and systems. Sometimes, however, one observes something that is infuriating. This anger is an indication that at some level you recognize that here is a problem that needs resolution. The gut feeling that the other person is wrong, or that there is a better way to do it, or that a product or theory is ugly or klunky, is a good guide to choosing an interesting topic for yourself.

Setting out with irrational determination to prove someone wrong provides a drive that can allow you to break out of your preconceptions. Such base emotions can be a strong creative force, causing you to dig deep and work intensely. The effort to refute someone can even lead to evidence supporting them or to a different topic altogether. Intensive rivalries, as in the race to discover DNA (Watson, 1968), can also provide this essential intensity. Thus whereas the finished product may appear dispassionate, truly creative work is often driven by strong passions.

Where there's smoke

A good strategy for finding an interesting problem is to follow the fire trucks, because "Where there's smoke there's fire." When there is intense debate on a topic, inconclusive or contradictory data, or terminological confusion, then things are probably ripe for a creative redefinition of the problem or application of a new method. If, however, your tendency is just to choose sides, then you are merely more kindling and should stay away from the fire. A creative redefinition comes about from recognizing how it is that each side in a debate could have come to their given conclusions. Once one can see clearly how the two sides could come to opposite conclusions from the same facts, then one is in a position to redefine the problem and create a resolution to it. Adler (1985) gives a beautiful illustration of this technique in his analysis of where Western philosophy has gone astray. He shows that modern philosophy has become irrelevant to the common (or even educated) person, which was not previously the case. Whereas philosophy in principle should be a useful guide to how to think about issues, about how to live and our place in the world, and about how to discover, its modern incarnation is none of these and is not even of much help in science. Adler identifies the crucial mistakes of thinkers such as Kant, Hume, and Locke, and how those mistakes created a mistaken path down which subsequent philosophers blindly followed. Once the initial mistake is uncovered, the subsequent elaboration of philosophical systems that lead to absurd conclusions (e.g. the relativistic view that scientific knowledge is purely a human construct, as though airplanes remain in the air by popular consensus) can be clearly seen and more coherent philosophies (which interestingly do provide guidance to the everyday person) can be developed (e.g. Adler's philosophical guidelines).

The Medawar zone

There is a general parabolic relationship between the difficulty of a problem and its likely payoff. Solving an easy problem has a low payoff, because it was well within reach and does not represent a real advance. Solving a very difficult problem may have a high payoff, but frequently will not pay at all because one is more likely to fail and because it may be difficult to take advantage of success. For example, designs for a tunnel under the English Channel were proposed over 100 years ago, but costs were prohibitive. The Greeks had an atomic theory over 2500 years ago, but had no way to test it. Many problems are difficult because the associated tools and technology are not advanced enough. For example, one may do a brilliant experiment but current theory may not be able to explain it. Or, conversely, a theory may remain untestable for many years. This is the case with some theories in physics at the moment (e.g. dark matter, string theory, cosmic inflation theory). Thus, the region of optimal benefit lies at an intermediate level of complexity, what I call the Medawar zone in reference to Sir Peter Medawar's (1967) characterization of science as the "art of the soluble." These intermediate level problems have the highest benefit per unit of effort because they are neither too simple to be useful nor too difficult to be solvable.

The issue of what is interesting and what is solvable lies at the heart of great discoveries and what we call genius. What is notable about great innovators is that they have an instinct for identifying this type of problem, and even when they are wrong, they are wrong in an interesting way and on an interesting topic. Some who choose to grapple with the big questions fail because they address problems not ripe for solution. The more common problem afflicts the average person who shies away from really interesting problems in favor of easier ones. In science, a focus on the easier, routine problems is a characteristic of T. S. Kuhn's "normal science" (Kuhn, 1970).

Working on too-easy problems is disadvantageous both because no one may notice your results (yawn!) and because "easy" or small problems often turn out to be merely pieces of a larger puzzle and only soluble in that context. For example, in the first X-ray pictures of DNA (Watson, 1968) two forms (A and B, differing by water content of the sample) of diffraction pattern were evident. James Watson and Francis Crick did not focus on explaining or interpreting this difference, but rather they focused on the more difficult problem of the DNA structure. When that puzzle was solved, the A and B patterns were easily interpreted.

When someone succeeds in frequently hitting the target (the Medawar zone), that person will often appear to be more intelligent than a pure IQ test would indicate. To an extent, the feel for interesting problems can be transmitted by contact, which explains why the institution of graduate-student-as-apprentice is so successful and why certain laboratories ferment with new ideas. Such labs are often observed to fade away or return to what is considered normal after the death or departure of the person or persons who provided the creative spark.

The creative spark is not easily obtainable through the formal textbook portion of professional training, and it may not arise spontaneously. For example, Richard Feynman (1984) recounts his experience as a visiting faculty member in Brazil in the 1960s. Physics in Brazil was just getting started. To outward appearances, the faculty knew the facts. Library and laboratory facilities were adequate. The students did very well on tests. Yet there was almost a complete lack of comprehension of the process of innovation and discovery. Science was a textbook exercise of learning definitions rather than one of discovery. For example, in a physics class that he was teaching, the students could define the properties of light polarization resulting from light striking flat surfaces at extreme angles, but could not give a single example even though they were overlooking the changing color of the sky's reflection from the ocean's surface. That is, they were unable to relate the definitions to the real world. Even in the United States today, entire departments or disciplines sometimes get stuck in such a listless or out of touch state.

Releasing creativity

Most people can learn to be far more creative than they are. Our school system emphasizes single correct answers and provides few opportunities for exploratory learning, problem solving, or innovation. Suddenly, when one becomes a graduate student or professional, however, it is expected that one is automatically an independent thinker and a creative problem solver. To a significant extent, creative responses are a matter of orientation and perception rather than a special skill or a dimension of personality. We may characterize noncreative responses as being stuck with a limited vision, an inability to see things in a novel way, a conventionality rut. I thus next focus on ways of encouraging creative approaches and reducing blocks to creativity. This aspect of creativity only addresses the ability to develop novel ideas or approaches. Other aspects of creativity involve the full elaboration and development of a novel idea into a finished product, a topic addressed in the following chapter.

Barriers to navigation

In the early fifteenth century, Prince Henry the Navigator of Portugal set out to explore Africa and open it to Portuguese trade (account in Boorstin, 1983). Portuguese expeditions began to work their way down the western coast, always within sight of land. Upon reaching Cape Bojador, a rocky stretch of desolate coast with treacherous currents, the Portuguese sailors would inevitably turn back, convinced that this was the end of land and that no ship would ever pass it. Prince Henry sent out 15 expeditions between 1424 and 1434 until finally one succeeded by sailing a few miles out to sea and going south for a few miles, thereby passing the dangerous rocks and currents.

As a navigation feat, this maneuver was trivial. While it is true that their ships were not strong and their navigation tools were primitive, the major barrier was not technological but the fear of losing sight of land. We can say that the feat of Columbus was far more difficult technically (not to get lost), but he too faced a major barrier of fear, both in his sponsors and in his crew. Once the Atlantic was crossed this fear was swept away and only the reasonable fears of shipwreck, scurvy, and sea serpents remained. Many barriers are of this type. An item becomes fixed in the mental landscape, immutable. What lies beyond the barrier becomes not merely unknown, but unimaginable. Major enhancements in creativity can be achieved by developing the courage to recognize and overcome mental barriers, just as the Portuguese and Spanish sailors did.

A simple test for creativity involves giving test subjects a set of objects and a goal, to see if they can use ordinary objects in unusual ways (e.g. a rock as a hammer). Noncreative individuals are often stumped by these tests. In the professions, too, objects become fixed in meaning. In many cases, an assumption comes to have the rock hardness and permanence of a fact.

My children had been playing with some yarn for months, calling it spaghetti for their toy kitchen. When my four-year-old daughter started twirling it around to the music, one piece in each hand like the Olympic gymnasts, my five-year-old daughter became upset because you do not twirl spaghetti around and dance with it. Similarly, it is often observed that young scientists or those venturing in from other fields often make the most revolutionary breaks with tradition: they are able to ask, "Is this really spaghetti?"

Those whom we note as outstandingly creative have often been described as possessing a childlike innocence or sense of wonder, and they ask seemingly naive questions. Einstein asking what would happen if you rode on a rocket at the speed of light and looked at a mirror held in front of you (he concluded that you would not be able to see your reflection), sounds like the ultimate naive question, like the silly questions kids ask, but it turns out to be one with profound consequences. This attitude contributes to creativity by keeping the mind flexible. Ambiguity and the unknown make many people nervous, however. It was not until the late fifteenth century that European map makers would leave sections of their maps empty. Before this period, they had filled the empty spaces of their maps with the Garden of Eden, the kingdoms of Gog and Magog, and imaginary peoples and geography (Boorstin, 1983). We do not easily suffer blank spaces on our mental maps, either.

A major obstacle to discovery is not ignorance but knowledge. Because Aristotle was so comprehensive, logical, and brilliant, his writings became the ultimate standard of truth for 2000 years. A major portion of Galileo's works was devoted to disproving Aristotle so that the reader would be able to grasp his arguments. The difficulty was that a single authority (Aristotle) was held in such high regard that alternative views could not get a hearing. In more recent times the work of Freud has had a similar effect. Freud's system of analysis posited certain mental constructs a priori such that it was very difficult to amend or improve his theories. It also focused on motivation (sexual drives) to the exclusion of process (how does one solve problems or process information), as noted by Arieti (1976). The result was that some psychoanalysts remained Freudians but many psychologists began to ignore Freud altogether in order to make progress in their work. B. F. Skinner's Behaviorism (Skinner, 1959) provided another classic barrier to knowledge. By defining behavior totally in terms of stimulus and response, and claiming that the mind was inherently unknowable, Skinner created a barrier to any understanding of phenomena such as emotions, memory, and cognition, subjects which in fact we are able to learn quite a bit about.

Another type of barrier of the mind is the definition by the professional community of what is a serious problem and what is not. Until the late 1970s, physicists regarded turbulence as largely beyond the terra firma of well-behaved phenomena subject to "real" scientific study. The discovery of the mathematics and physics of chaos (chaotic attractors, universality, relations to fractals, and all the rest) is rightly called a revolution (Gleick, 1987b) because it brought within the realm of orderly study an entire class of phenomena previously classified as "void, and without form."

In the case of chaos, there was a well-defined phenomenon, turbulence, that was deemed intractable. Another common situation is when a topic is initially not even recognized as such. Mandelbrot's breakthrough in the "discovery" of fractals was in recognizing a common set of properties in what were previously unconnected mathematical oddities. When Darwin found earthworms interesting enough to write a book about them (Darwin, 1881), the world of science was quite surprised. Recognizing problems that others do not even see can be considered a prime characteristic of the truly innovative.

Barriers to recognizing a phenomenon or problem are many, including concreteness, visualizability, and complexity. Before Riemann, the geometry of Euclid was identified with the three dimensions and properties of our sensory world. The axioms therefore were too concrete for anyone to conceive of altering them. Breaking this concreteness barrier led to many forms of non-Euclidean geometry, a result that was later instrumental in the development of relativity.

Visualizability can also be a limiting factor. Once Poincaré sections of the orbits of strange attractors were published, it became evident to everyone that there was some kind of regularity to turbulent (chaotic) phenomena. Formal proofs of this fact were far less influential to the general scientific community because they are much less accessible (Gleick, 1987a).

Complexity and heterogeneity are also major barriers to recognizing problems. The innovation of Newton was in recognizing that a ball thrown in the air and a planet circling the sun are "the same" with respect to gravity. He made the further crucial abstraction of treating his objects as point masses, reducing complexity to a minimum. These abstractions and simplifications of Newton are, in reality, simple, but only after the fact.

It is characteristic of mental barriers that once overcome they are never given a second thought. The Spanish navigators never considered the Atlantic a serious problem once it was crossed. Of course, many scientific achievements really are complex. The mathematics necessary to grasp quantum mechanics is quite difficult and is not just a mental barrier. Nevertheless, one must always be alert for barriers that can be circumvented.

A significant barrier to navigation is the set of structures we have erected to facilitate our work: namely, academic departments. The current system seeks to fill all the square holes with square pegs. The biology department wants one geneticist, one physiologist, and one ecologist, but they don't want three generalists who work in all three areas. In what department would one put Darwin: genetics, geology, taxonomy, or ecology? Darwin considered himself to be a geologist, but the world remembers largely his biology. Should Goethe be in the literature, biology, physics, or philosophy department? He actually was most proud of his work on optics, though that work was largely flawed. Would Newton or Fisher find comfortable academic niches today? The current rigid departmental system is confining to the truly creative person and discourages the vitally important cross-fertilization of models, data, techniques, and concepts between disciplines.

Don't fall in love

A scientist should never fall in love – at least not with his own theories. A scientist should keep an open mind (though perhaps not so open that his brains fall out if he bends over). But do they? In fact, scientists often subscribe to a particular framework or support a particular theory in opposition to others. A scientist may support or oppose string theory or the big bang, may favor controlled field experiments or models of ecosystems. The question is, does he defend his preferred theory or framework tooth and nail, or allow for the possibility that flaws may exist?

In manufacturing, many companies have embraced continuous improvement. This is the process of continuously reducing cost, increasing product durability and customer satisfaction, reducing waste and pollution, etc. For any given topic in science, it should be possible to improve experimental procedures, develop better statistical tests, explain anomalies, etc. Falling in love with a theory and developing that contented glow of companionship is a guaranteed way to overlook opportunities to improve the field. To put it another way, the scientist, like a craftsman, needs to develop an eye for imperfections which can be improved. If different individuals seem to debate without resolution, perhaps it is because they are using terms differently (e.g. "neurosis" or "personality") or perhaps it is because certain phenomena act differently in different systems (e.g. "limiting factors" in a desert versus a rain forest). In either case, an opportunity exists for clarification. There are always odd phenomena that don't fit in, and they offer opportunities for discovery. For example, certain lizards can reproduce without males. How did this evolve? What about mutations in such a system? Maybe the opportunities exist at the experimental level. For example, by implanting radio-frequency identification (RFID) tags under the skin of fish or radio collars on deer, the animals can be repeatedly located without further disturbance.

A theory is not your friend or your spouse. "I'm OK, you're OK" is not a good framework for you and your theory. You should never be the defender of the faith or keeper of the holy relics. Even your own personal theory (often likened to one's child) should be subject to merciless interrogation. If your theory is attacked, you should examine the criticisms. If they have merit, revise the theory. If not, perhaps you need to explain it better or deal with objections explicitly or collect better data.

One should also not fall in love with your own status as an expert. It is satisfying to be recognized as knowledgeable. If you write a book on a topic it can be very tempting to view this book as a crystallization of what is known and correct, and oneself as the arbiter of disputes. But if you are the authority and your past work is perfect, how can you bring yourself to improve it?

If you shouldn't fall in love with your theories and your own greatness, how can you love your work? The real excitement of science is the thrill of the hunt. It is the discovery of contradictions and the solving of puzzles. It is outsmarting the prey. This is the secret to staying fresh.

Don't be an expert

All graduate students are taught that it is essential to become an expert. As a short-term goal this is, of course, valid. Academic search committees are also looking for experts. As a lifestyle, however, becoming an expert can inhibit creativity.

Why is this? After all, it seems that an expert has more tools at his or her disposal for solving problems. The problem revolves around our mental constructs. In learning a subject, we create a network of facts, assumptions, and models. Once we think we understand something, it is linked up to an explanation and supporting ideas. This construct may not be true, but it comes to seem real nevertheless. As one becomes more of an expert, a larger and more complex network of facts and explanations accumulates and solidifies, making it difficult to entertain radical alternative ideas or to recognize new problems.

The expert is in danger of developing the small cage habit. Zoo animals, when moved to a larger cage, may continue to pace about an area the size and shape of their old smaller cage (Biondi, 1980). An Aristotle or Freud may create a set of bars within which most people pace rigidly, never noticing clues from outside the cage. The danger in becoming an expert is that one tends to build one's own cage out of the certainties and facts which one gradually comes to know. Dogmatism builds cages in which the dogmatic then live and expect everyone else to live also.

How does one not become an expert? Astrophysicist S. Chandrasekhar gave a remarkable television interview a few years ago. He has led a scientific career notable for a rate of productivity that has not slowed down at all into his seventies. When asked how he has avoided the drop in creativity and productivity that plagues many scientists, he replied that approximately every seven years he takes up a new topic. He found that he would run out of new ideas after working in an area for too long. This pattern led him to tackle such topics as the dynamics of stellar systems, white dwarfs, relativity, and radiative transfer. Although all these subjects are in astrophysics, they are different enough to present unique problems.

We need only turn to Darwin to find a truly remarkable example of the value of changing topics. He wrote books on the origin of coral atolls, the geology of South America, pollination of orchids, ecology of earthworms, evolution, human emotions, the taxonomy of the world's barnacles, and movement in plants. Although most of this work in some sense related to or led toward his grand project (evolution), they were all quite different in themselves. When he decided that a topic was interesting, he would delve into it in depth for a period of years, write up his results, and move on. After his early books on geology, he only returned to the topic a few times during the remainder of his career. In today's atmosphere, he would have been encouraged to follow up on his early study of corals or geology for the rest of his career. Imagine him in a modern geology department telling his department head that he planned to spend the next 20 years working on evolution, earthworms, and orchids.

It is easy to protest that learning a new subject is too hard and takes too long. I am not suggesting that everyone can or should strive for the diversity of Charles Darwin. Taking up new subjects within a discipline or linking up with related disciplines appears more difficult, however, than in fact it is. It is much less difficult than the original university experience, because the mature professional has an arsenal of tools, terms, and techniques that are transferable between topics. I assert that the value of cross-fertilization far outweighs the cost of learning new skills and facts. Studies have shown that a wide spectrum of interests is typical of highly creative scientists and helps account for their creativity (Simonton, 1988).

Practical problems beset the brave soul who eschews the expert label, however. Getting grants for research in a new area will be difficult. Department heads will frown. Within many corporations one may place one's career at risk. Exploring new territory inevitably evokes the Columbus response: shaking of heads and muttering as you disappear over the horizon and a hero's welcome when (if) you return. A strategy some researchers employ is to maintain a home base of expertise in a narrow area to keep department heads and deans happy, with frequent excursions to diverse topics to stay fresh.

Don't read the literature

Graduate students are inevitably told to read the literature to get started. This advice is fine for students, because they are used to looking up the answers in the back of the book anyway and repeating the examples they have seen. For the practicing professional, however, this first step can be inhibiting. First, it channels your thoughts too much into well-worn grooves. Second, a germ of an idea can easily seem insignificant in comparison to finished studies. Third, the sheer volume of material to read may intimidate you into abandoning any work in a new area. Medawar (1967) also advises against reading too much, arguing that study can be a substitute for productive work.

My recommendation for the first step (after getting the germ of an idea) is to put your feet up on the desk and stare out the window. Try to elaborate the idea as much as possible. Do some calculations or quick lab experiments. Write a few pages or sketch out a design. Only after the idea has incubated and developed will it be robust enough to compare it to existing literature. Given a certain level of knowledge in a subject, you know generally what is going on, so you are not likely to be reinventing the wheel. When you go to the literature, you may find that someone has preempted you or that your idea is invalid, but at the risk of only a few days or weeks of work. The cost of good ideas killed off too soon is much higher than the cost of some wasted effort.

Visualization

People differ in the extent to which they use visualization as a problem solving tool. Einstein held visualization to be a key mental tool. Others favor pure mathematics or verbal models. Communication of scientific results is critically affected by the visual tools used, such as maps, plots, charts, and diagrams. Understanding of a phenomenon can even be inhibited until a key visual representation is invented. There is no formal training provided to scientists in either internal use of visualization or usage of formal model or data presentation techniques.

The internal use of visual imagery is not communicated in published work and is only brought to light when famous scientists are interviewed about their work habits or write biographical material. It is thus difficult to assess how universal visualization is as a problem solving tool.

Visualization can help one develop an intuition for phenomena as well as aiding in theorizing. Einstein famously pictured himself on a rocket traveling through space when contemplating inertial frames and relativity. Visualizing manifolds and movement across them can provide a complement to working with the formal mathematics of dynamical systems.

Sometimes successful visualization is itself the solution to a problem. As the study of proteins and enzymes has progressed it has become clear that their 3-D geometry is the key to their function. Their shape enables them to block a site or bring two compounds together to enhance their reaction, for example. Computer tools that can represent protein structure, including animations of their changes in shape, have revealed a great deal about how they work.

Maps are critical to representing many processes and types of data in geology, ecology, conservation biology, climatology, pollution control, hydrology, and other fields. They can help to illuminate patterns and clarify causation. At the same time, because of their high level of abstraction they can also mislead. For example, a map depicting different plant communities on a landscape for purposes of evaluating biodiversity is subject to distortions due to resolution, color scheme, and classification level (Loehle and Wein, 1994). Since colors chosen to represent a plant community (itself an arbitrary classification) are arbitrary, each map created by a different person can look entirely different. There is no correct answer. This is true even with quantitative data such as temperature, where a different color gradient can give different impressions of which regions are unusually warm or cold. When any map is made, some minimum cell size must be chosen for drawing the map. As resolution becomes coarser (say from $100\,m \times 100\,m$ to $10\,km \times 10\,km$ cells), smaller scale units must either be averaged somehow or the majority type within the larger cells will determine the type for that cell. This means that rare classification units will vanish in large-scale maps. Conversely, at very fine scales the units in question may not be defined. For example, at a 1 m resolution it is not meaningful to define "forest type." The level of classification can also lead to confusion. At a very broad scale, we might classify only forest versus grassland, and the landscape will look very simple. As finer gradations of these categories are defined, the map starts to look more complex. What is the "true" level of habitat complexity? There is no correct answer to this question.

Line graphs and bar charts are the standard visual problem-solving tools in science. They are more familiar to most scientists but can still easily mislead. For example, on bar charts some measure of uncertainty should be shown, but even when it is it may not quite represent the statistical test that the reader assumes. That is, the reader may assume that if two bars have error bars that overlap then they are not different, but the proper statistical test may not exactly match this intuitive understanding. Log or other transformations of data before plotting have consequences that readers may not appreciate. Similarly, overlaying plots in different units can be misleading, as can the choice of vertical scale which can exaggerate or minimize the perception of change.

Sometimes a theory is embodied in a diagram or figure. Such figures can still be misinterpreted. For example, fractals are defined in terms of a repeating geometric construction, such as branches branching off of branches or curdling in 3-D space. A fractal can lead to power-law type relationships, but it is also possible to get such relationships from other dynamic processes. It has also become clear that many real objects that appear fractal may not be strictly self-similar; that is, their dimensionality changes with scale. As another example, catastrophe theory is a topological theory of the stable points of a dynamical system. Certain criteria must be met for a system to follow these dynamics. In many applications of the theory, the cusp manifold has been applied to systems that do not meet these criteria (cf. Loehle, 1989). In these cases, the manifold is being used to explain something that is only vaguely similar, rather than structurally similar.

Visualization is thus a potentially powerful tool for scientific problem solving and communication, but one that is not without pitfalls.

Work habits

Work habits are a crucial component of strategic creativity. Many creative people are unable to follow through on an idea. Others are less productive than they could be or complain of distractions. This section provides tips on work habits in the context of enhancing creativity.

Slow down

Given the pressures to get grants, run a lab, and write papers, the life of a scientist can become frenetic. Many become too busy to return calls or answer e-mail. They can be constantly on the go. It may thus seem paradoxical to suggest that greater productivity can be achieved by slowing down. My suggestion is rooted in two objectives that are served by slowing down: (1) thinking deeply about a problem to avoid heading off in the wrong direction and (2) pausing to allow the partially conscious mind to ruminate on a problem.

In comparisons of student problem solving (Whimbey and Whimbey, 1976), it was thought that the better students would be found to read a difficult problem faster and solve it faster. In fact, the good students took much longer to read the problem, because they were thinking about it, but then took less time to answer the questions or do the math. The poor students often were jumping ahead and solving the wrong problem. On simple problems, there was little difference in performance. This habit of jumping ahead also leads too often in technical areas to solving the wrong problem. The pace of professional life has become so frenetic that activity and motion have come to replace thought. The need for careful thought and planning is particularly acute for projects involving complex systems such as large-scale software projects, integrated manufacturing, large construction projects, and high-tech product development. There is a simple test for freneticism: merely ask someone, "Why are you doing this task?" If they are too busy to answer or cannot explain it, the ratio of thought to activity is too low.

The practice of rumination is critical for thinking about a partially completed project, not just for the early theoretical stages. When a manuscript is in pretty good shape, it is critical to slow down. Don't rush it into print. Develop the habit of leaving it on your desk while you ponder it, mull it over, contemplate it, chew on it, turn it over in your mind. Did I do the right statistical test? What objections will reviewers raise? Did I explain myself well enough? What did I miss? These questions are crucial and can make or break a paper.

An effective technique, good for deeper contemplation, is the walk. This technique is looked down on today as being too low-tech. Besides, someone walking is obviously not working. Darwin used to take an hour's walk every day around a course he had laid out. He would become engrossed in his thoughts; therefore he put some small stones at the start, kicking one off at each round so that he did not have to keep track of how many circuits he had made or worry about time. It was during these walks that he wrestled with the deepest questions.

The practice of taking long walks as an active part of intellectual activity used to be a common part of academic life in Europe. Professors would take their graduate students on walks to debate, discuss, and question. These days, graduate students are lucky to even see their professor in the halls. Our idea of a walk is going to the copy machine. Some psychologists have found that taking patients for a walk is very effective in getting them to open up and express themselves. With our short attention spans these days, it would no doubt require practice to be able to come to conclusions or formulate complex thoughts while walking and remember them back in the office, but it can be done and would be beneficial.

Think inside the box

The advice to "Think Outside the Box" is so ubiquitous that it is even used to sell tacos ("Think Outside the Bun")! When this phrase was coined, it might perhaps have been that people were too conventional, too stuck in routine ways of thinking. Today, however, with instant messaging and nonstop entertainment, it is not that people lack divergent thinking, but rather that they lack the ability to focus. In some jobs this may not matter much, but in science and engineering this is a recipe for failure. This is true for several reasons.

Science consists not just of a method, but of a body of knowledge and of technique. The established body of textbook knowledge forms a set of constraints (a box, if you will) that the scientist must work within. Any study must obey the conservation of momentum, of energy, and of mass, whether you like it or not. Ignoring physical constraints can lead to failure. There are also constraints consisting of the established methods, procedures, and statistics for a field. These methods have been widely tested and debated over time and accepted as reliable or at least standardized. For example, for DNA studies there are lab procedures to guard against cross-contamination. If you ignore this guidance you may end up sequencing your own DNA instead of that of the target organism.

Staying within the box has other benefits. There are many types of studies where there is some latitude in method. For example, various methods can be used to sample amphibians in the field. Only when different studies use the same method, however, can their results be compared or combined into a larger-scale analysis. Using established methods and statistical procedures can also assist with communication, since readers will more readily understand what was done, and reduce the chances of rejection by reviewers.

When individuals are unaware of the box, hold innovation to be supreme, or are impatient with such tedious issues, they may attempt to circumvent scientific method. For example, claims about health benefits of vitamins and herbal remedies are usually based on anecdotal evidence or just a certainty that they *must* work. If one really needed all the items on offer at the health food store, there would not be room in one's diet for food, just supplements. In less extreme cases, impatience with mundane details of "the box" leads to sloppiness. In science, "Thinking Outside the Box" leads to bad procedure and quackery.

There is another sense in which thinking inside the box is crucial in science. Science is not just about generating *new* knowledge (innovation) but about generating *reliable* knowledge. Flashes of insight are often wrong. In order to conduct a research program, it is first necessary to define a plan of action. It is then necessary to execute that plan precisely. Being spontaneous is very nice for taking a drive on Sunday, but if you keep changing your mind about what to do and how to do it during the course of a study it will require a miracle for the end result to make sense.

In addition to careful execution of the study plan, attention must be paid to precision in carrying out the individual steps. If there is an established procedure, it must be followed. If field water samples must be chilled right away after collection, leaving them out in the sun is not being "creative": it is being sloppy. If your instruments need daily calibration, do it. If you need to randomize the location of rat cages to avoid position effects, do it. There is nothing worse that doing an entire study and then realizing that you have instrumental drift or that your treatments are confounded. Don't leave reviewers and critics with easy "Yes, but..." opportunities. The practice of science is thus a paradox that not everyone can adjust to. The box that one must work within may feel confining to the artistic temperament. If you accept the box too completely, you will be unable to discover anything new. If you ignore the box, reviewers will reject your work. The goal is to transcend the game, to be so good at the rules (given knowledge) and so precise with the accepted procedures that you become free to discover truly new things with reliability.

If you can't walk, try running

I have been a recreational jogger for 30 years. I sometimes find that a pain in my ankle that I feel when walking or jogging will go away if I switch to a sprint. This cure suggests a strategy to overcome writer's block (designer's block, etc.), which afflicts many scientists. The scenario I often observe is that someone finishes a project or gets a new idea and then sits down to "write up the results" but can't get started. It reminds me of the Peanuts comic strip in which Snoopy is trying to write a great novel and keeps getting stuck on "It was a dark and stormy night."

Starting at the first word to write up an entire document is rather intimidating. The walking writer, like Snoopy, is noticing the pain in his ankle at every sentence and is likely to stop and massage each sore spot, thus repeatedly getting stuck. Such jerky motion is also anathema to creative thought. Sprinting can sometimes cure both problems. One should sit down with a cup of coffee and define a short piece to be written in a defined interval, say the methods section in one hour. The introduction is not where one should start; rather it should be last, after the technical parts are all worked out. Then sprint without worrying about grammar or style, which can be corrected later. Leave blanks where the references and figures should go. Often this approach will get one off the mark and writing may continue for several hours. If it turns out not to be a good day, the sprinting technique at least allows for an hour or two of solid work. The utility of this approach depends on the style of the professional and is most useful for hyperactive individuals who do not like to sit still and for perfectionists like Snoopy who get stuck on the first sentence.

The four hour work day

A frequent cause of inadequate creative and professional performance is mental fatigue caused by excessive pressure, hours, or effort. Many people wish to be perceived as hard working, so they do everything with intensity and put in long hours. Unfortunately, this can make the brain rather fuzzy, which is detrimental to strategic thinking. Strategic thinking is not a brute force approach, but rather requires wise and efficient use of mental faculties to increase effectiveness. The logic is that it is more important to be effective than to be busy, because in the realm of complex problems the most important thing is to not start off down the wrong path. An army of programmers can be busy as bees developing a new program, but if the basic design for the software is faulty, the result will be of no use to anyone. It is for this reason that I recommend the four hour work day.

Few of us can work at full capacity, thinking clearly and profoundly, for the duration of every work day any more than we can run at top speed for the same distance that we can jog. This is the myth of the modern work environment, that anything can be expected of workers and they must deliver. Most people fake it, and will not admit when they are at less than 100%. The result is that terrible decisions are made, experiments are ruined, and serious bugs are introduced by people who are fatigued. Ironically, working too hard can lead to the need to work even longer in order to fix the goofs that result. In contrast, it is much more likely that one can work at 100% mental clarity for about four hours. If one keeps this in mind, then a distinction can be made between critical issues that need full clarity and intense effort, which become part of the four hours of work per day, and those parts of a project that are routine and become part of the rest of the day. That is, if you only expect yourself to be brilliant for four hours per day, then you may actually live up to this standard. During the rest of the day, there are plenty of routine tasks to accomplish such as returning calls, coding a clearly designed subroutine, ordering equipment, attending seminars, editing reports, etc. We may again turn to Darwin as an example. He habitually spent about four hours writing, which he found exhausting because this was the creative part of his work day, and spent the rest of the day writing correspondence, doing experiments, reading articles, etc. Ralph Waldo Emerson also spent half a day writing, and then spent the rest of the day in his garden or doing other chores. In the case of Emerson, we may be confident that while gardening he was simultaneously engaged in rather profound thought, and was thus not really goofing off. A key to this approach is to choose that part of the day for the most difficult work when one is at peak alertness. If you are a morning person, try to teach your classes in the afternoon. Don't wait until your 3 p.m. low-energy spot to sit down and get to work on your big theory.

To reiterate, the value of the four hour work day is that certain components of creative and technical projects are more crucial than others and thus require more intelligence to do properly. In science, the choice of problem and approach for solving it are absolutely crucial to success, and require one's full attention. In architecture, the overall concept for a building is the crucial step, with much of the drafting requiring skill but not brilliance. In computer science, the functional specification for a piece of software and the basic layout, flowcharts, and algorithms are the crucial steps which require extra care. For these crucial steps, if one is not thinking 100% clearly, then one is likely to introduce bugs that require significant effort to fix. In all of these cases, if the difficult parts are not done well, subsequent work on the details is a waste of time. Thus the recognition of which components of a project require extra care and attention is a critical component of ultimate effectiveness. Proper use of this approach can make one look much smarter and more efficient that one actually is, because so little time will be wasted and because such difficult problems will be solved.

Be unrealistic

It is a fatal mistake to have a realistic estimation of your mental capacities. Someone who is realistic will never attempt problems that seem hard, because few of us are Newtons. On the other hand, creativity is only marginally related to IQ. That is, above a certain minimal level, IQ and college grades are not predictive of productivity, success, or innovation (Arieti, 1976; Gardner, 1983; Simonton, 1988).

As we look back on great scientific discoveries, many of them seem childishly simple to us now. The great innovation of Galileo was to avoid trying to explain *why* objects fall (as Aristotle had) in favor of quantifying *how* they fall. When Newton treated objects as point masses it was brilliant, but in retrospect it is a simple concept. The great innovation of Vesalius was to do dissections himself and base his anatomy book on what he actually saw rather than on the authority of Galen (Boorstin, 1983). His further innovation was to use medical diagrams in his book. All of these are elementary ideas.

Some may despair that all the easy ideas have been found, but this is far from true. In the last two decades, fractals and chaos have transformed the foundations of science, yet the basic concepts and even some of the formal math are intuitively obvious and simple once learned. Often the solution we seek will turn out to be simple and well within the reach of our intelligence. It is puzzling why discovery is so hard when the final result can often be demonstrated to an eighth grade class.

Let us be more explicit on this point. At any given time it will seem that all the good ideas have been discovered. The head of the U.S. patent office in the mid 1800s suggested that it be closed down because all of the good ideas had already been discovered (though this may be an urban legend), but the flow of new inventions and scientific discoveries does not seem to be letting up even now. Even in the realm of business, innovation does not stand still. IBM was the obvious giant in the computer field and without serious competitors, but nevertheless personal computers, work stations, distributed computing, and parallel computing originated elsewhere and brought IBM to its knees in the 1980s and after. Sears was the dominant department store in the United States for decades but missed the boat on several retailing innovations and was bought by K-Mart. Ironically, K-Mart had sold many stores during its own bankruptcy and had a pile of cash on hand. Ownership of Sears by K-Mart has done nothing to revive Sears. Impossible things like radial keratonomy for correcting nearsightedness, superconductors, worldwide e-mail, cell phones, movie special effects, bucky balls (60-carbon molecular cages), and fractals have all burst upon the scene in recent decades out of nowhere. The fact is that there are always surprises around every corner. Even in areas where it seems that everyone has come to agreement, there can be surprises. In astronomy as of this writing, the age of the universe seems to be less than the age of some nearby galaxies. This is certainly a surprise and shows that we clearly are missing something somewhere. The well-known "fact" that cholesterol causes heart attacks by clogging arteries has recently been challenged by a new theory that iron is the actual culprit. By this theory excess iron (which the body accumulates over years) catalyzes cholesterol into a form that causes the damage to arteries. This theory jibes with the fact that men with very high iron levels have higher heart attack rates and explains why women have an increased risk after menopause (they stop losing iron in blood during menstruation). This new theory is not proven, but shows how a well-accepted theory (bad cholesterol) may turn out to be completely mistaken. It also illustrates how simple a revolutionary theory can be and still be new. When we look at any domain, the same result holds: new results, ways of doing business, inventions, designs, and innovations are all around us. The realistic person "knows" he isn't smart enough to discover or invent these things. The strategic thinker realizes that he doesn't need to be so smart, if he is clever.

Bandwagons and parades

Science is the search for reliable knowledge, but knowledge of what? Why study stars or DNA or dinosaurs? Well, many scientists study what they are interested in … but this just begs the question, doesn't it? What makes one topic interesting or important? There is no true answer to this question. Individuals study certain topics because they find them interesting in the same way that people like baseball or Roman history. As with any topic, there are fads in science. These fads can affect funding and the ability to get recognition for one's work.

Consider the war on cancer. Clearly here there is a goal that all would agree to. The bandwagon aspect of this topic is that all sorts of studies are justified by their implied benefits in this war. But this distorts priorities because many basic aspects of cell biology and human physiology are *not* going to lead immediately to a cure, but it is necessary to build up this knowledge base to be successful. Funding can thus be channeled into all sorts of magic bullet studies, most of which fail to pan out.

Fads put the individual researcher on the horns of a dilemma. To follow every fad as it comes along is to be aimless. To eschew all fads is to risk not getting funding or recognition. There is no perfect resolution for this conundrum, which I leave for the reader to ponder.

Inverse procrastination

The first priority of the innovator is procrastination. Only by putting off routine duties and avoiding committee assignments can one find time to daydream and browse in the literature. I do not believe it is fair to call this procrastination and avoidance *irresponsible behavior*. Rather, it has to do with lead times being more important than deadlines. The gestation time for ideas, methods, and models is often quite long. The Eureka! phenomenon is usually the tail end of a long process of puzzling over a problem, reading about it, and discussing it with colleagues.

For example, ever since my teens I have been fascinated with the ability of some trees to live for thousands of years. I read accounts of tree life span and counted rings on stumps without any goal in mind for many years. But eventually this information led me to a new approach to the problem of the energetic costs of achieving great age (Loehle, 1988b).

I believe that most creative professionals have a long list, or zoo, if you will, of perhaps only partially articulated questions and puzzles that they mull over and that guide them. The need to feed the inmates of this zoo at regular intervals is strong, because these ideas will blossom into the next set of research problems. This drive leads to what I call "The First Law of Inverse Procrastination": always put off some of what you should be doing today so you can do something that might be relevant later.

Surfing

If I say that creative work is like surfing, you will think I am from California. By this analogy, however, I mean that good ideas come sporadically and unpredictably and should be pursued as they pass by, just as the surfer pursues the wave. Some waves are small, some large. Some days the surf is up, and some days it is not. For the really big waves, it can take real effort to stay on the crest. The little waves can be caught by jotting down notes wherever you are. When the surf is up, it is crucial to recognize it and, like the California hot-dogger, cut classes if necessary to hang ten. At such times, one should shut the door and disconnect the phone. In such creative waves, sometimes entire first drafts of papers can be written in a continuous burst or entire systems designed in one session. Authors have written entire books in a few weeks when inspired in this way. Edward de Bono wrote *Future Positive* in eight days and *The Bridges of Madison County* was written by Robert James Waller in three weeks. Such work is often of the highest quality even though hurriedly done.

Does such an approach mean one should be a prima donna, only working when the mood strikes? Certainly not. On days that are not good for surfing, there are articles to read, manuscripts to revise, equipment to order, papers to review, phone calls, meetings, and so on and on. The point is not to be moody but to be receptive to the creative muse (to be musey, if you will). Designating a fixed time of day for technical work or following too rigid a pattern of work is detrimental to such receptivity.

Surfing applies to topics popping up, as well as to being inspired in general. To cite B. F. Skinner (1959), "a first principle not formally recognized by scientific methodologists: when you run onto something interesting, drop everything else and study it."

This principle points out a fundamental problem with the current peer review grant-giving process as well as with project scheduling for engineering R&D. The current review (or planning) system requires one to lay out, in some detail, the steps and procedures one is going to follow through several years and what the expected outcome is going to be. The U.S. Department of Energy actually sent out a memo to its labs around 1990 (which I saw) requesting that it be notified at least six months in advance of any major discoveries! This demand is completely unrealistic, because research is a contingent process. Rigid scheduling also precludes following up interesting leads. Examining Faraday's notebooks, one sees that he did several experiments per day in an iterative, tinkering type of research. How could he have planned this research in advance or presented it to a review panel? In the context of engineering, it is clear from the examination of the history of the development of any product or technology (see Petroski, 1992) that iterative tinkering is far more the rule than is following a plan set up at the beginning of the project.

Today's highly competitive climate has led to the misconception that the quality of proposed work and its outcome is predictable from a detailed proposal. Few if any really surprising discoveries get explicitly funded this way. As Koestler (1964) noted, "The history of discovery is full of arrivals at unexpected destinations, and arrivals at the right destination by the wrong boat." A much better practice is to fund investigators, as does the Howard Hughes Medical Foundation, for three- to five-year periods based on the individual's track record rather than to fund a detailed proposal. This practice frees up the truly productive from the huge overhead of chasing grants (as much as 50% of one's time) and from making overly rigid research plans. One can not predict or control what the creative person will do, but he or she can be encouraged by adequate support.

Intensity

Certain types of creative tasks require exceptional powers of concentration, a certain intensity. This is most true of those types of tasks requiring strategic thinking. Tasks in which each short piece of work stands alone may be done in spite of interruptions. When I am opening my mail it is only a slight distraction if someone drops by my office or the phone rings, but in the middle of writing a complex computer algorithm I do not at all enjoy being interrupted. Such an interruption can in fact cause a bug to be introduced that requires hours to fix.

Let us clarify this point. An airline pilot must concentrate on his task, but if he is interrupted or distracted for a moment the overall task is not at risk as long as he is not landing the plane at that moment. The unique aspect of creative endeavors is that complex mental constructs must be created and held in the mind for a task to be successful. An artist must generate a vision of the desired work and then hold this vision in mind while working. Sketches help make the vision concrete. Because the artist must hold this vision in mind while working, most artists require privacy and long periods without interruption. Many also refuse to allow anyone to view a piece before it is finished. Since the viewer does not "see" the painting the artist has in mind, they may react to incomplete design elements and thereby express emotions or opinions that are inappropriate or negative. This may cause the artist to react to these reactions instead of following his original vision.

In many creative and technical fields this same situation exists. An intense focus is needed to maintain a sufficient level of concentration to follow long trains of reasoning, to build up complex networks of relationships, to respond to subtle clues, and to envision complex structures. Without intense concentration, subtle and half-formed thoughts will slip away. We may in fact say that without intense concentration the benefits of taking long contemplative walks or of putting the feet up on the desk will not be realized.

Some years ago, in my work for a statistical software firm, I was debugging a very long piece of complex code written by other people. This bug was sort of a phantom that only occurred sometimes and was therefore driving us nuts. I was tracing through the program code line by line and simultaneously following the machine instructions on the debugger. This meant that I had to trace through thousands of lines of code and keep in mind the entire flow of control and many values in the computer memory. During three attempts, each of which took three hours or more, I had gone a little too fast or missed something, and had failed to catch the phantom in action. On my fourth attempt I was very careful and did not move from my chair or think of anything else for three hours, at which point I trapped the bug: I caught it executing a simple FORTRAN statement such as $X = Y$, but the machine executed a WRITE op code. This meant that we had a hardware bug in our VAX computer, which it turned out was in the disk drive controller software. This explained why the bug usually appeared after lunch: at this time we were usually working the machine pretty hard, and the disk drive reader software would get lost when working at high speed. Only intense concentration enabled me to find this bug.

Examples abound where intense concentration is essential to success. Before computers were available, Milankovitch postulated that the tilt of the Earth's axis and the precession of its orbit could lead to sufficient changes in light interception by the northern latitudes that ice ages could result. He spent many long months doing the calculations by hand. Any mistake could jeopardize the entire calculation. Kepler before him performed a similar computational feat to demonstrate that the planets follow elliptical orbits. The philosopher Eric Hoffer, almost blind, would compose his works while walking, and then dictate them later.

We may in fact generally say that the types of problems of concern in this book require intense sustained work characterized by complete concentration. Any problem that is highly technical involves synthesis of information, involves complex structures, or has multiple components falls in this class. It is simply not possible to jump from task to task when the problem is complex. A more formal analysis is presented in the following sections, but at this point we can simply observe that interruptions of a complex task introduce errors whose seriousness increases with the complexity of the task.

These observations bear on several misconceptions about problem solving. A major focus of "creativity" workshops is to get people to lighten up and be spontaneous. Much effort is focused on brainstorming and allowing absurd or novel ideas to surface. While all of this is certainly useful for people who are overly rigid and afraid to entertain new ideas, this focus on spontaneity is seriously misplaced. It creates the impression that leaping about and quickly generating clever ideas is sufficient for success; that is, it completely ignores the issue of follow-through. In fact, given half a chance most people can generate lots of clever ideas, but lack the ability to follow through on complex problems that require sustained effort, long chains of reasoning, and concentration. Truly innovative work requires both periods of generating new ideas and periods of sustained work to follow through on the initial good idea. Unless the problem is trivial, brainstorming is only the first step.

A second implication concerns work habits. I mentioned earlier that good work requires periods free from interruptions. Such solitary work is difficult for many people. Those who are very gregarious feel the need to talk with others about their ideas or to call a meeting about it, but this mode of work does not lead to sufficient intensity or to the generation of sufficiently long trains of thought to solve very complex problems. In fact, complex problems have seldom if ever been solved in a meeting or by a committee. The focus on busyness and visibility in many companies means that everyone is almost always interrupted before a complex task can be completed. If someone went off to a remote site for six months to succeed in solving a problem alone, they would come back to find they were no longer considered a player, and were no longer part of the power system. This is part of the reason why many companies are so short-sighted in their planning.

Understanding the key role of intensity and concentration allows us to grasp the peculiar phenomenon of the overachiever. I remember in high school that following IQ testing, some students were identified as being "overachievers." With only an average IQ, they were doing exceptionally well in school. If we believe that IQ tests measure anything at all, and they are said to be quite good at predicting success in school, then overachievement on this very measure of what IQ is supposed to predict would seem to be nonsensical. The simple explanation is that these individuals work with greater than average intensity. This extra focus and concentration lead to greater performance on complex tasks than expected from performance on short-answer tests such as an IQ test (see following sections for more details). This same factor can also explain underachievement in many cases. To complete real work, even a genius must concentrate, but this is hard work and requires discipline. Without sufficient intensity, complex problems (involved in any type of high achievement) can not be completed successfully, even by a genius.

How does one foster this critical intensity? It has been observed that when poor students (in both senses) join a good chess club after school, their grades go up. Attempts to find specific cognitive skills fostered by chess that might transfer to school work have failed. This increased performance could be attributed to a change in attitude, but I believe it is at least in part due to the students learning how to concentrate intensely. It is not possible to play chess without concentrating, and it thus fosters this particular skill. The same may be said of mathematics and computer programming, which also are not possible unless one concentrates. As noted, performance on all complex tasks requires this kind of intense concentration, and thus the payoff from chess and similar activities can be substantial. In general, activities that promote intense concentration, particularly if this concentration is mental rather than attentional (one may be an intense fisherman, but this type of concentration is not useful for strategic thinking problems) will provide valuable training and the skill acquired will carry over to other domains of problem solving.

Conclusions

The path of creativity is strewn with the bones of those consumed by the vultures of mediocrity, accountability, and responsibility. One can not schedule creative breakthroughs, budget for them, or prove them in advance to a review panel or manager. An entirely different, flexible approach to discovery is necessary to encourage creativity. The concept that time is too valuable for staring out the window or reading for pleasure is equivalent to riding a bicycle under water. Free and undirected thought and research are essential.

On the other hand, one must not live in the creative moment permanently. The imagination is very powerful and can easily mislead (Arieti, 1976). One whose ideas remain conceptual will never know if they are valid, feasible, complete, or useful. They will remain in the realm of dreams and play. One's goal as a strategist is to produce finished products of some sort. This means that innovative ideas must somehow be converted into concrete form. Dreams must be converted into designs, inventions, systems, or some other product. Thus creativity per se is only the first step. What is necessary to be an effective innovator is for the process of converting ideas into the realm of action to be facilitated and made efficient. To do this, one must understand the entire problem solving process, not just the step of generating novelty. The next several sections discuss the concrete steps of converting ideas into reality. First, one must take a vague initial concept or intuition and turn it into a substantive discovery. This process involves elaboration of an initially vague idea so that it becomes operational, measurable, and can be communicated. Next, one must convert a discovery into a product, such as a piece of software, or a theory, or an experiment, or a manufacturing system. Finally, one must check the new theory or product against reality to verify that it is useful, efficient, novel, or correct.

2.2   DISCOVERY AS A PROCESS

Introduction

At the heart of strategic thinking is the discovery of new ways of solving problems, new ways of organizing information, and new ways of designing products and systems. Thus the discovery process needs exploration here. The previous section on creativity set the stage for an examination of the discovery process by setting out the approaches and attitudes that make discovery possible. In this section the nature of discovery and innovation are explored and myths about them cleared up. The presentation of discovery will be in the context of scientific discovery, because this is the only field in which discoveries can be verified by others. In other contexts, a discovery (e.g. how to best organize a corporation, or a new art form) is difficult to verify, and in the end can only be objectively tested by in fact applying scientific methods. The examination of discovery in the context of science, where we can examine the component steps out in the open and verify our analysis of the discovery process, enables us to learn how this process works and then apply this process reliably to other fields of endeavor.

Scientific discovery is one of the most dramatic and exciting products of the human mind and ultimately is the source of our advanced technology. There is tremendous pressure to increase the rate of discovery as more and more key societal issues are seen to contain a scientific component (e.g. AIDS, global climate change, tropical deforestation). Currently expectations of what science can deliver are very high because of science's past breakthroughs, but the general perception of how science works is not in harmony with its actual operation. This discrepancy leads to the attitudes reflected in AIDS protests in the United States that the government is not doing enough, even though over $1 billion is being spent annually on research, which is more per AIDS mortality case than is spent for any other disease. The expectation that money alone is required to order a cure on demand results from a serious misunderstanding of the discovery process. This same misunderstanding is evident in corporations which expect to be able to order up new products from their R&D departments with the snap of a finger, and become impatient with research when progress is slow.

A clearer understanding of how discoveries are actually made is necessary, so that barriers to discovery can be removed and discoveries can be rewarded and enhanced. As Bauer (1992) pointed out, science is not based on a fixed recipe or method (the "scientific method"). If it was, we would not be frustrated by problems such as cancer, and scientific progress would be predictable. Rather, science proceeds by a process similar to solving a puzzle. That is, discovery is a process, not an event, and in fact is a rather complicated process. The official version of how science is or should be done conflicts with this reality. Discovery is in fact best achieved by the application of strategic problem solving techniques, as will be shown.

Some persistent myths about scientific discovery cause a great deal of trouble and confusion in the practice and public perception of science today. Three great myths, which contribute to problematic public attitudes and form a sort of triumvirate of misunderstanding, are the Eureka! myth, the hypothesis myth, and the measurement myth. These myths are prevalent among scientists as well as among observers of science, and spill over into attitudes and approaches prevalent among those of the public who wish to appear logical and scientific. The Eureka! myth asserts that discovery occurs as a flash of insight, and as such is not subject to investigation. This leads to the perception that discovery or deriving a hypothesis is a moment or event rather than a process. Such events are singular and not subject to analysis. The Eureka! myth has tainted the entire field of creativity research and training. The hypothesis myth asserts that proper science is motivated by testing hypotheses, and that if something is not experimentally testable then it is not scientific. The fact is that many of the great breakthroughs in science involve methods and not hypotheses (e.g. the electron microscope, X-ray diffraction methods, polymerase chain reaction (PCR) technology; Hall, 1992; Crease, 1992) or arise from largely descriptive studies (e.g. evolution, plate tectonics). Furthermore, "testable" and "experiment" are commonly taken to mean a simple dose-response or analysis of variance type test, which is far too confining for complex theories. The third myth is the measurement myth, which assumes that the objects we can observe are the real variables in the system, which may be interpreted unambiguously. This is a concrete view of the world in which gross national product, consumer confidence, food web, and ecosystem are assumed to be real objects that we may measure. Following a detailed characterization of these myths, I present an alternative, developmental model of science.

Myth 1: the Eureka! myth

It has long been assumed that the philosophy of science may be helpful in epistemology or in the logic of deduction, axiomatization, or justification, but that it has little to say concerning the process of discovery. Popper (1963), for instance, stated that discovery (the generation of hypotheses) is not subject to formal rules, so that neither inductive nor deductive logic are reliable (though they are sometimes useful) paths to discovery. This idea has led to the view that scientific discovery is somehow a mysterious subconscious process, a conclusion that does not necessarily follow from Popper's argument. Many have written about flashes of insight, dreams, the psychology of discovery, etc. (e.g. Root-Bernstein, 1989). In this view, discoveries appear as a spontaneous act, in a flash. If one is not a "creative type," then one won't have creative flashes.

The view that discoveries, insights, or hypotheses are obtained in a flash is what I call the Eureka! myth. This is the myth that discoveries are made all of a piece, and consist of an event or moment or thing. It is reinforced by dramatic accounts like Kekulé's discovery of chemical ring structures in a dream and similar accounts (cf. Hadamard, 1949). Conscious work–incubation–insight has in fact become the popular image of discovery, as documented by Langley and Jones (1988). Popular courses on creativity focus almost exclusively on such Eureka! moments. Such colorful, flash-of-insight accounts capture the imagination and are certainly interesting; however, they may be misleading. They draw attention to only the dramatic cases and suggest that the process of discovery is not subject to study or dissection. They also draw attention away from valid insights or new ideas that require nurturing, development, and maturation. That is, if it isn't completed in 10 minutes, then forget it. Discovery as instant oatmeal.

In contrast to this Eureka! view, I would like to argue for the central role of pattern recognition in the discovery process. The human brain is wired for pattern recognition, as Margolis (1987) has argued at length. This process is neither inductive nor deductive, nor is it rule based, though it can be mimicked by rules in some cases. For example, everyone can recognize, at a flash, hundreds if not thousands of faces, without using statistical hypothesis testing or deductive logic. Many people are equally good at musical patterns, being able to recognize thousands or even tens of thousands of songs. They generally would be unable to do so from the sheet music and absolutely unable to do so from a sonogram.

Besides visual and auditory patterns, some people have a facility for recognizing other, more abstract types of patterns. Einstein described his own thought processes as being of this type (Hadamard, 1949). Such a skill is particularly useful for finding relationships in phenomena that do not necessarily have a visual representation.

There is a crucial difference between detecting patterns in science and the popular concept of creativity. Creativity workshops focus on spontaneity, on being uninhibited, on play, on humor, and on dressing up in a gorilla costume. But scientific discovery is not a free, uninhibited, artistic type process. It is far closer to puzzle solving or mechanical work. That is, a pattern or mental structure or understanding does not necessarily come all as a piece and in a flash, but rather is built up slowly and piecemeal as one links facts together and builds and rearranges a mental framework for the problem. It involves tinkering, puttering, patience, and stubbornness. That is, we may say that the scientist is involved in constructing patterns. These patterns consist of networks of relationships between facts, assumptions, mathematical relations and methods, measurement techniques, rules of thumb, and hunches. A "discovery" involves an expansion, rearrangement, or simplification of all or part of this network of relationships. A discovery is thus not a *thing* or *event* and rarely involves only a single step.

It is necessary to clarify here that there are exceptions that occur when a thing is actually discovered, as in being found. For example, Columbus discovered the Americas. We may discover a new species of plant or a new moon around Jupiter. However, often we speak of a discovery such as the discovery of a new medicine or vaccine or of a fundamental atomic particle. These "discoveries" usually involve a long series of steps and are constructed piece by piece. It is not possible to look in the microscope and "discover" a vaccine; much work in DNA analysis, protein characterization, and cell biology goes into this discovery, and in fact there is no single moment when the "discovery" can be said to have occurred. Similarly, physicists usually don't just find a bunch of Higgs bosons lying around on the laboratory floor and then run out and announce a discovery. Usually there is a lot of work that goes into such a discovery, and a lot of analysis. Thus the discovery of *things* such as moons or new species might more properly be called findings or observations and should not be confused with other types of discoveries.

Myth 2: the hypothesis myth

It has become a truism that one can not design a proper experiment without a clearly stated hypothesis. To an extent this is a reasonable approach as a response to the pure empiricist tradition, because the mere gathering and cataloging of "facts" only leads to a very large pile of facts and very little knowledge. The emphasis on hypotheses may be traced to Popper's (1963) Principle of Demarcation, which gives testability as the demarcation between scientific and unscientific (e.g. astrology, Marxism) theories. This perfectly correct argument has been carried to too fine a level of detail, however, by the general scientific establishment.

The argument that a *theory* must be testable has been extended to the requirement that every aspect of science must involve a theory (hypothesis) and a test of that theory (hypothesis). This narrow interpretation of "testability" is blind to the fact that much of science is really technology. Scientists spend much of their time developing instruments, software, and methods to measure and detect phenomena (Crease, 1992; Hall, 1992). This activity is science; however, it is not a theory nor a test of a theory. In addition, empirical, descriptive studies are a necessary part of science. It is from them that sufficient data may be obtained to allow a pattern (structure, relationship) to be detected, possibly forming the basis for further study. Such preliminary observed patterns are not hypotheses in the sense of Popper because they are empirical rather than explanatory.

We may trace this confusion over what is a hypothesis partially to the excessively simplified presentations of the scientific method taught in school and partially to the confounding of concepts between a scientific hypothesis and a statistical hypothesis. A *statistical hypothesis* is a probability statement in terms of outcomes. It concerns measurable events and magnitudes. In contrast, we may have a perfectly valid *scientific hypothesis* (theory) but be unable to specify its outcomes. A deductive step is required to go from the scientific hypothesis to the statistical hypothesis, and this step is not always easy or clear cut. After it was proposed that hadrons are composed of quarks, it took much effort to determine how to test this idea. Superstring theory remains untestable at present. In contrast, if we conduct a purely empirical study of the efficacy of a pesticide (with no biochemical theory of how it works) by measuring dose versus kill rate, we have no scientific hypothesis about causation, but we do have a statistical hypothesis framed in terms of the experimental design. The way that statistics is taught in college contributes to these difficulties. The distinction between a scientific and a statistical hypothesis is rarely made. Statistical tests are presented with the assumption that the question is either obvious (difference in mean income between two groups) or already determined. No mention is made of how one derives a testable and useful hypothesis from a theory. In fact, statistical examples in texts are almost never in the context of testing or refining a theory; rather they are almost all purely empirical relations or questions (finding defective light bulbs, age versus income, comparing fertilizer effects). By making statistics a separate academic discipline, it has become cut off from science and the students exposed to it learn it out of context and assume that *this* is what "experimental design" is. In practice, scientists learn the proper role of statistics as they see how statistics is used by more experienced scientists, and not from their statistics courses.

A final source of confusion arises from multiple definitions of "scientific hypothesis." If I say "I think X and Y may be related in some way," this is a hunch or intuitive guess. This is what we have when we *think* we have found a regularity (pattern) in nature. Comparing X and Y does not "test" this guess because being wrong on such a hunch or "working hypothesis" does not cause any change in a scientific paradigm. In contrast, a scientific hypothesis (theory) is based on cause–effect reasoning. A scientific hypothesis does not merely state that X and Y may be related, but it *explains* why they are related. Should we find that X and Y are *not* related when a theory predicts them to be, then something is wrong with the theory. The correlation between body size and life span in mammals is not a hypothesis or a theory, but any potential explanations for this correlation are.

The essential point about hypotheses is that there are different types and levels of hypothesis. At the first stage of discovery we have a hunch or intuition that we have detected a pattern or relationship in nature. This working hypothesis is a perfectly valid subject of study. A working hypothesis may lead to the discovery of an empirical relationship, which may be quite precise but is not a theory. A working hypothesis becomes a scientific hypothesis or theory when one offers an explanation for a pattern or relationship in terms of cause–effect or structural properties. A statistical hypothesis may be purely empirical (dog food brand A is better), or descriptive (a quadratic line fits these data), or it may be used to test a scientific hypothesis after suitable deduction and definition of variables.

Myth 3: the measurement myth

The measurement myth is the assumption that the fundamental objects or variables (primitives) that are our objects of study have real existence or are unambiguous. That is, once we define objects or variables and use them in discourse we come to assume that they are real and that we can measure or study them; this is reification. We then base our statistical (experimental) analysis on the measures of these objects and assume that our analysis bears on the real world. For example, the concepts food web and ecosystem seem concrete and are used as if they are concrete, but in fact they are fraught with observational, boundary, resolution, and other difficulties and ambiguities. Similarly "gene" as commonly used really refers to the one gene one trait assumption, which is often not valid (there is no "gene" for producing an arm or the eye). We may also recall gross national product, consumer price index, and consumer confidence as variables that are often treated as if they were concrete and objective properties of the real world, which they are not. In reality we rarely make discoveries on the basis of clearly defined variables, nor are the meanings of any such variables as we do use necessarily fixed and unambiguous. Pattern recognition violates the assumption that the objects to be measured are initially well defined or even identifiable. This fact becomes clear when we examine pattern recognition in some everyday contexts. It is not at all obvious which features we use for recognizing a familiar face. We never measure facial features to identify people and would not even be able to say which variables are informative nor how to measure them. Furthermore, we have no clue about how we are processing this information. The same ambiguous use of information is a fundamental aspect of scientific discovery.

Initially, a discovery is like the recognition of a face; the discoverer feels sure that he has found a pattern, but at first this pattern is intuitive. The elaboration process must make this intuitive pattern explicit, which means putting it in terms of measurements and data. As a discovery unfolds and is elaborated, the meanings of previously unobjectionable data (facts, metrics, variables, objects) may be called into question. The solid atom may turn out to be composed of parts. Fixed continents may turn out to move. Rulers may shrink and grow and clocks move at different speeds. Thus, the basic assumption of hypothesis-testing science that the variables to be measured are a priori well defined and quantifiable does not hold up in the early stages of scientific discovery. The path from pattern to theory to experiment is not necessarily straightforward.

It is stated by some philosophers of science that there are no "facts" separate from theory. From the above we can see the sense in which this is true. As a theory develops, it impacts and changes our concepts and our interpretation of facts. Quantum physics alters our concept of "empty space." The modern concept of wealth in economics includes informational and quality of life types of wealth that were not considered wealth in the 1800s. It is also unfortunately the case, however, that the implication drawn from this is that it is impossible to truly test a theory because it is all of a piece. In reality, a theory does not alter all of our definitions or perceptions of facts, only certain ones. There are generally many facts, metrics, or observables we may use to test a theory that are not influenced by it.

A new model

I would like to propose an alternative to the standard model of the scientific method. My model emphasizes pattern and its elaboration as being prior to formal statistics and hypothesis testing. The four steps are the following:

1. A pattern is found, a defect in an existing pattern is found, or no pattern is found where one was expected. The pattern may be in data, or it may be a meta-pattern, relating abstract concepts.
2. Elaboration takes place. A vocabulary must often be developed to describe the pattern. Methods of measurement (of shape or form, statistical properties, instruments, laboratory protocol) may need to be developed. Characteristics of the pattern are refined.
3. A theory or explanation for the observed pattern is proposed. The theory may need elaboration, rearrangement, or analysis before it yields testable predictions.
4. Experimental, statistical testing of hypotheses begins.

The usual concept of discovery compresses stages 1–3 into a product (a hypothesis or theory) with the process being hidden. During the conduct of routine science, one can often move rapidly from stage 1 to stage 4, and the processes of stages 1 through 3 tend to be glossed over or lost in the final scientific report of a study. In many cases, however, the real work of discovery takes place iteratively at stages 1, 2, and 3, but differs in nature and is often done by different people at different stages. That is, "discovery" really involves this entire process of iterative refinement, elaboration, and explanation of patterns perceived in nature. The same argument may be made about invention, which is typically viewed as an insight or aha! experience, but actually usually involves a prolonged elaboration and refinement process. That is, the original idea for the invention is not the invention itself, which may in fact take years of fiddling and refinement before it is worthy of a patent.

In the initial stages of pattern detection, it may be clear that a pattern has been found but not at all clear how to proceed next. Strange attractors were not well received at first when expressed mathematically (Gleick, 1987a), but when they were presented visually it was recognized instantly that real patterns were involved. For example, the sequence in time of water dripping from a faucet appears random, but when the dripping is plotted as an attractor (Gleick, 1987b), the pattern becomes apparent. Once attractors were recognized from graphic plots, it became clear that a descriptive vocabulary was needed. For example, the patterns of chaotic attractors are clearly not random, but are very difficult to describe. Statistically demonstrating the difference between random and chaotic time series is quite difficult and new methods have been needed (e.g. Tsonis and Elsner, 1992). The initial intuitive recognition of attractors has thus been augmented by the development of tools such as the concepts of scaling and universality and the use of Poincaré sections, though the descriptive vocabulary for attractors is still deficient.

We might recognize patterns via the Gestalt process of figure-ground reversal. For example, we might fail to see a consistent pattern in locations where a species is found but see a pattern in locations where the species is *not* found. As another example, one might study not only the attractor of a dynamic system but its repeller, which may provide new insights (Sidorowich, 1992). We might make a discovery by observing the lack of a pattern or response where one was expected. Such was the case in the discovery of symmetry breaking in physics (for example, in the generation of the asymmetry that produces a magnet).

Finally, we may make a discovery by observing a defect in a pattern, an anomaly (Lightman and Gingerich, 1992). Such was the origin of relativity: a defect, largely ignored, in Newtonian mechanics. This idea is revisited in later sections.

The process of pattern elaboration is central to the study of patterns for which cause and effect cannot be discerned immediately by experimentation. During the elaboration stage, the problem of how and what to measure is crucial. For example, in the early X-ray diffraction studies of DNA (Watson, 1968) what was being measured and how to interpret the data were not at all clear. What is foreground and what background, what is data and what noise, all need to be worked out. In addition, a vocabulary for description is essential. The trained taxonomist need only glance at most trees to identify them, but in doubtful cases he must refer to a key that uses a specialized descriptive vocabulary for leaf shape, flower parts, etc. Much of mathematics serves to provide a descriptive vocabulary. The process of developing a vocabulary in itself may change the perceived pattern. For example, in describing body form in the context of taxonomy, it was found that axis of symmetry was a useful descriptor. Radial symmetry is characteristic of simpler, more primitive forms (such as sea urchins) compared with the bilateral symmetry exhibited by all vertebrates. The recognition of bilateral symmetry led to the question of symmetry breaking such as the handedness of the large claw in crabs or right-hand dominance in humans. Thus, the process of elaboration for both description and measurement may change our perception of the original pattern or uncover further patterns, all without necessarily involving hypotheses, explanations, or statistical tests. Further aspects of pattern elaboration are discussed in Loehle (1988a).

Discussion

The human brain not only recognizes patterns, but it seeks them out and will generate them even when they do not exist, as in clouds or ink blots. We may characterize much of science as the identification of complex natural patterns; their elaboration, quantification, and explanation; and, finally and definitely last, their statistical testing. The admonition that real science requires that one begin with a hypothesis and then test it is the death knell of discovery, largely limiting the practitioner to questions such as "the effect of levels of X on Y." Clinical trials, drug safety testing, agricultural research, and other fields are held in a death grip by this formalism. For example, simple-minded linear extrapolation of dose-response experiments currently governs our food, drug, and contamination regulations but totally ignores the complex machinery within cells that combats low levels of oxidative damage and that repairs DNA. This is not to say that experimentation is bad, far from it, but simple dose-response experimentation, whether for fertilizer, dog food, or carcinogens, is totally incapable of refining a vague theory, or of truly testing how things work. No single experiment or test will clarify the functioning of the immune system, the processes governing ice age cycles, or the factors regulating evolution. Such topics require complex and abstract theories which, because they are complex, require a focus on their internal structure and their development.

Another implication of the present argument concerns the lack of progress evident in certain fields of science. Bauer (1992) notes, for example, that in the social sciences conflicting theories and schools of thought exist without any resolution in favor of one or the other for indefinite periods of time. He attributes this problem to the ineffectiveness of the "reality therapy" that other sciences must endure. I would in contrast argue that lack of progress in such fields is due to failure of the elaboration step. As Hall (1992) and Crease (1992) pointed out, technique is a crucial aspect of scientific progress because it allows us to measure things such that hypotheses become testable. In the social sciences there is no shortage of "scientific method" or hypotheses or advanced statistics (Bauer, 1992), or even of data, rather what is missing is the capacity to develop a theory's internal structure and to reliably measure things or to define entities which are measurable. The type of elaboration of technique described by Hall (1992) and Crease (1992), such as development of X-ray diffraction technology or radioisotope tracers, often stops at the preliminary stage in the social sciences. Concepts are proposed which might be important or causative, but the further steps of elaboration are often not successful. How do we reliably measure alienation, social unrest, happiness, or culture? Such terms illustrate the fallacy of reification, that just because one can use a word the word therefore refers to a thing that exists. Instruments such as questionnaires and tests have been developed, but it is not clear what they actually measure. For example, an IQ test measures something that predicts success in school, but it is not clear that this something is "intelligence." Theories in the social sciences tend to be built of such concepts that we seem to understand but can not precisely quantify, and as such are subject to multiple interpretations. In spite of the use of good scientific method, vague concepts interfere with rigorous testing against reality. This is not to say that true insights are not obtained in these fields, but rather that it is impossible to eliminate beliefs that do not accord with reality because tests that are acceptable to all parties as a proof cannot be obtained when the basic concepts are vague. In ecology we can point to similar problems that result from the use of concepts such as ecosystem, fitness, and biodiversity that are also ambiguous. As an antidote, I suggest that more effort be expended to develop theories that are based on the mechanisms of operation in the system. For example, cognitive psychology focuses on how people process information and make decisions, with a good linkage between theory and experiment. In economics we can point to studies that ask what behaviors will result from a system consisting of imperfect decision makers operating on the basis of incomplete and even erroneous information. This approach is much closer to reality than the standard analytical models of omniscient utility maximizers.

We should not underestimate the danger posed by the myths of the scientific method. For example, Robert Gallo, the discoverer of the AIDS virus, was accused of scientific misconduct for, among other things, supposedly taking an AIDS culture from a French scientist and claiming it was his own. In fact, cross-contamination turned out to be the likely culprit. David Baltimore's team was accused of fraud and misconduct for what, in the end, turned out to be sloppy record keeping. Both men were cleared of misconduct but only after immense cost to their careers. The treatments of the Gallo and Baltimore fraud cases suggest that the auditors expected the type of rigor and formality in these studies that might be found in a bank ledger, on the assumption that the scientists were (or should have been) following a set procedure. In reality, scientists on the cutting edge are struggling to measure things that no one understands, using methods that no one can prove are right. No one can see the bacteria or virus on a finger that results in cross-contamination; there is not necessarily a bad guy. Similarly, inventors struggle to create something they can't define by methods they can't defend. Aside from outright fabrication, the types of errors being criticized in the Gallo and Baltimore cases can be found in any laboratory where really innovative work is being done. Legitimate arguments about lab technique, artifacts, and interpretation of data can occur in any field and for any experiment. Similarly, the expectation that any drug or vaccine can ever be declared totally safe results from this same misconception. Clarity, certainty, and well-defined lab protocols become available only long after the discovery phase. In fact, when we reach this stage the work is commonly turned over to technicians, as are AIDS blood testing or soil fertility testing today. During the early days of the

AIDS epidemic, for example, there was a complete muddle concerning symptoms and causation. Measuring the level of the HIV virus and correlating that to the degree of symptom expression gives a negative result because of the long latency of the virus and its odd behavior. Similar confusion exists now concerning chronic fatigue syndrome.

Much confusion and misdirection results from the mistaken idea that first and foremost the scientific method is about doing controlled experiments with statistical rigor. In fact, it is largely about the struggle to convert an intuitively perceived pattern into something sharp and definite. By the time experiments are feasible and standard statistics apply, the battle is mostly won.

Once we understand, as above, how discovery operates in science, we may apply this approach to enhancing discovery in other domains as well. For example, the techniques of brainstorming commonly promoted for business problem solving are clearly inadequate in the context of strategic thinking. Brainstorming may be adequate for naming a new product or discovering features people would like to see added to a car, but it is too superficial to address complex problems because it fails completely to assess the pattern elaboration stage, in which we uncover the complexities and subtleties inherent in our original insight. For example, brainstorming would not have helped Beethoven develop his unique and sophisticated musical styles, nor would it be of much help in developing a complex software system. The same may be said for the standard model of discovery in psychology: analysis–incubation–insight. This model assumes that one pass at the target is sufficient to achieve a final result. In fact, any really original design or problem solution requires elaboration, iteration, and evolution. Both brainstorming and the standard model of discovery neglect any mention of testing whether the insight or solution one has achieved has any relation to reality. That is, hypothesis testing is not incorporated into these techniques. Thus we can not say that these techniques are reliable in general or even useful for complex problems in particular. Strategic thinking, in contrast, considers all of the mentioned solution steps specifically in the context of complex problems.

2.3 STRATEGIC PROBLEM SOLVING

Introduction

In this section I wish to discuss the problem solving process itself, not in terms of the mental operations of thinking about a problem but in terms of the actual actions one takes to obtain a finished product, whether that product be a design, a plan, an experiment, an invention, or a piece of software. In particular, I build here on the discussion of discovery in the previous section and put discovery in the context of the practice of solving real problems and producing technical products. The discussion here is where the material in the first two chapters is put into practice as components of an overall model of practical problem solving.

The structure of a problem has a great deal to do with the best way to organize one's work. For cleaning up the house, almost any order of doing the work will be equally efficient, but for complex tasks the way work is done has a significant bearing on success and productivity. In particular, a strategic problem solving approach becomes crucial in this context to avoid a high risk of failure or a low level of productivity. The risk of failure for complex projects is not trivial. Gibbs (1994), for example, documents very large project delays (up to twice as long as planned) and large probabilities (up to 50%) of project cancellation for large software projects (over 10 000 function points). Many of these failures result from a lack of knowledge about the inherent differences between simple and complex problems and how they must be approached and managed. An analysis of multistep problems shows how such failures as noted for software projects are a natural feature of their size, how failure risk can be assessed, and how complex projects can be approached to minimize failure and increase productivity. This same type of analysis provides a guide to the organization of work on any technical or professional project, from writing software to construction project management to inventing. This analysis is presented next. While the following few pages entail a little more heavy reading than the rest of this book, they provide key tools for understanding the structure of complex problems and for understanding large-scale failures and fiascoes. The reader should therefore stick with it. Nothing more advanced than a little multiplication is required to get through it.

Problem solving as a multistep process

In considering the problem solving process, I wish to distinguish three basic types of problems: parallel, sequential, and iterative. In parallel problems, steps are independent of one another. For example, for a set of simple addition problems, each problem is independent. One's score thus represents the successes divided by the total tries. If one is skilled enough to get 90% of the problems right, then on a long enough test one's score should be a 90. Most tests in school are parallel in nature.

In a sequential problem, the success of the whole depends on each part or step. The classic example is electronic components in series. If any single component fails, the whole device stops operating. If each component has a 0.1% chance of failure during some time interval and 20 components in series make up a critical device, then the probability of failure of that device is $1.0 - 0.99920 = 0.02$, which is unacceptable for a critical device such as those aboard a space shuttle. Even with a component failure rate of 0.0001, we get a 0.2% failure rate for the device, which is still unacceptable, especially considering the thousands of devices and systems of devices involved. The only solution is redundancy, the course in fact taken for rocket systems. The probability that n redundant components will fail simultaneously is the nth power of their single failure rates. For the example above with a 0.1% chance of failure per component, if even two redundant units are used for each of 20 components in series, the risk of failure goes down to $(1.0 - (0.001)^2)^{20} = 0.00002$. Alternatively, if we have two entire redundant systems of 20 components each, but with no redundant parts internally, we have $(1.0 - 0.999^{20})^2 = 0.00039$, which is still a very low failure rate.

Iterative problems have quite different properties. Examples include writing and editing a document such as this one, computer programming, and painting. Iterative problems have the property that work previously completed can be fixed, reworked, and improved. No one has ever written more than a few hundred lines of computer code without errors, but the errors can be fixed iteratively without starting over each time.

Technical and planning tasks consist of problems of all three types. Proof reading a report for mistakes is largely parallel; the errors found are independent of one another. Complex calculations are highly sequential, but they are typically repeated to check the answer. Laboratory work is also often highly sequential. Writing a paper, theorizing, and data analysis tend to be highly iterative.

The basic premise of my model is that much research, invention, and design work is a multistage, multistep process, where stages are qualitatively different sets of related tasks (e.g. theorizing versus experimentation) that themselves consist of discrete tasks (steps). A discovery or end product such as a patent, device, software system, or publication is not a single "thing." It is rarely complete with the initial "Eureka," but rather it represents a cumulative series of steps or solved problems, all (most) of which must be correct. Newton had to invent calculus before addressing mechanics, a stage that itself consisted of many steps. In a typical experiment one must have an idea; design, set up, and execute the experiment; perform statistical analyses; interpret the results; properly compare these results with the literature; and then communicate the results to an audience. An inventor must generate and elaborate a concept, create a prototype, test it, refine it, and document the result. Individual steps, such as performing transplant surgery on laboratory animals or observing lion behavior in the field, have a significant risk of failure. Failure at any step certainly leads to lost time and effort, thereby reducing productivity, and it may destroy the entire experiment. (I deal with creative use of failures below.) A certain level of failure (or even reworking) may use up all the time and money on a particular project, leading to total failure or the waste of a large block of time. Even when the specific component tasks in a technical project are parallel or iterative, the overall success of the project depends on most of them being done correctly. For example, a finished manuscript can be rejected because of poor writing, improper statistics, inadequate experimental technique, or other causes. Only a single serious flaw is needed to make publication unlikely. In addition, even a delay can be equivalent to failure if someone else makes a discovery or brings a product to market first.

These effects can be quantified by using standard critical path analysis models applied to problem solving. The simplest critical path (failure analysis) model where all steps are in series is simply Psuccess = p^n, where p is the probability of success (assumed constant) for a single step, and n is the number of steps. For a problem that consists of 10 steps, the probability of success is p^{10}, where p is the probability of doing each step correctly. Two individuals could have an "A" average in college with 90% and 97% correct answers on tests. However, for a 10-step sequential problem, $0.90^{10} = 0.349$, whereas $0.97^{10} = 0.737$, more than a twofold difference in the chance of success for any given project (and thus in productivity). For a 20-step problem, $p = 0.9$ gives only a 12% chance of overall success, while $p = 0.97$ gives a 54% chance of success. This simple calculation does not tell the full story, of course, and I develop a more complete model next, but this calculation does point out something that has been overlooked in the past: small differences in problem solving skill lead to widely divergent results for multistep problems. This may help explain why top students in lower grades suddenly become average students as material gets more difficult and long problems become the rule.

A critical path model of productivity

At this point, I would like to introduce a more complete simulation of multistep problem solving. Consider the office of a professional as a factory whose products are inventions, designs, computer systems, plans, proposals, patents, discoveries, or publications. I wish to model first the small "factory" with perhaps a few assistants, but in which most of the ideas, analyses, and writing and much of the laboratory work is done by the professional per se. We can see that the "raw materials" of this factory are not really mice or chemicals but concepts, hypotheses, experimental skill, base knowledge, computing skills, etc. There are many ways to model this factory. In a study of a furniture craftsman's shop, we might distinguish design, wood working, assembly, and finishing stages to partition labor costs and find bottlenecks. In my model of the professional as factory, I have defined three basic stages: ideation (I), execution (E), and communication (C). Each stage consists of multiple steps. Success in a step of any of these stages is given by p, which represents problem solving skill. We are interested in the number of successful projects (productions) completed in a period of time. The products of completed projects are publications, patents, software, machines designed, etc. The model is a mix of sequential and iterative components. In operations research terms, the model is a queuing model. Jobs (productions) queue up for the attention of the single operator. Each job is processed through three work stations (I, E, and C). Multiple jobs can be in process, but the operator can only work on one at a time. Errors are randomly introduced but are not detected until the specific checkpoint (defined below). Errors lead to reworking, thereby tying up the operator (the professional). Thus, no simple equation defines the level of productivity. Rather, a simulation of the process is necessary. The simulation model is defined as follows:

1. The probability of success at each step is p, a measure of technical skill. In the simulation, uniform random (0,1) numbers (R) are drawn and compared to determine success at each step. For $R > p$, the step is a failure. Each step counts as 1 time increment.
2. Stage 1 (ideation) has I steps; all must be completed successfully. Failure leads to the loss of all time required for both stage 1 and stage 2. That is, a failure means that a useless experiment or design project was performed because of a nonfruitful idea.
3. Stage 2 (execution) has E steps. If a step is unsuccessful, the model is reset to the beginning of stage 2, representing a mistake in an experiment, product development, or design project that is based

on a basically sound idea. The mistake leads to repeating the execution of the project (an iterative action). The entire time required for stage 2 is lost by failure, on the assumption that errors in the execution stage are not detected until the end of that stage.

4.   Stage 3 (communication) has C steps. Unsuccessful steps are repeated (as in rewriting a paper). Complete failure of communication is not allowed because of the iterative nature of writing. (One can always rewrite a document.) Allowing failure at stage 3 would accentuate the nonlinearities and differences between short and long productions discussed below.

Although projects are assumed to vary in length, the average project is modeled. Few-step versus multistep projects are modeled explicitly below. Since the scientific method allows for some experimental error, "failure" in this context refers, for example, to ruined or uninterpretable experiments, not just to occasional measurement error; or to a useless software system, not just one with a few bugs. A lifetime of professional output was simulated by setting $T = 8750$ time steps. For a 35-year, full-time career, this equates to one real work day per simulated step. One day is a reasonable estimate for a discrete piece of an overall project, such as writing a section of a paper or analyzing data. The overall number of projects completed over a career as a function of p is given in Figure 2.1. The upper line is for simpler

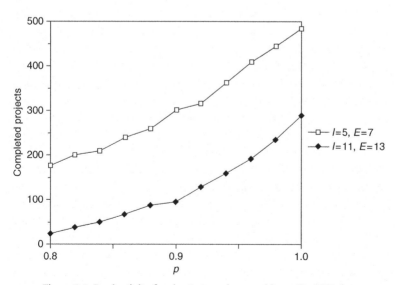

Figure 2.1 Productivity for short versus long problems. $T = 8750$ time steps

Table 2.1 *Lifetime productivities predicted by the model for long problems*
(I = 11, E = 13, C = 6)

	Full-time researcher	Full time with assistants	Quarter time (full teaching load)
Brilliant scientist ($p = 1.0$)	291	582–1000	73
Average scientist ($p = 0.8$)	24	48	6

problems with few steps ($I = 5$, $E = 7$, $C = 6$). The lower line is for more difficult problems ($I = 11$, $E = 13$, $C = 6$). These parameter values were chosen on the basis of my experience, but they could instead be empirically determined. The maximum possible total productivity (at $p = 1.0$) is of course lower for the problems with more steps.

The output of the model, shown in Figure 2.1, corresponds fairly well with reality. The maximum possible lifetime productivities (for $p = 1.0$) shown are in the range of that observed in real life (486 for short problems, 291 for long problems). For skilled scientists who must teach full time with only summers free, this would give lifetime productivities of 121 and 73 for short and long productions, respectively. Again, these numbers are in accord with reality. Employment of technicians and students and coauthorship practices can effectively double or triple these figures, giving maximum lifetime productivities that closely match all but the most prolific (Table 2.1). The most prolific scientists overall are usually heads of large teams (Anderson, 1992), in which the average number of papers per worker may be only 1–3 per year (total papers divided by total lab staff). For example, Boxenbaum *et al.* (1987) showed a strong negative correlation between percentage of first-authored papers and total number of publications. Publication records for some of the most famous, working largely alone and on difficult problems, include Einstein (248), Galton (227), Binet (277), James (307), Freud (330), and Maslow (165) (Albert, 1975). However, much of these authors' productivity was in the form of books, so their true productivity was higher than these figures indicate. An example of what can be accomplished by a single person working largely on smaller (though still difficult) problems is the case of I. J. Good of the Statistics Department at Virginia Tech, with a lifetime total of over 1600 publications (pers. comm.), many of them notes or commentaries. Other examples include John Cairns, Jr. (Biology

Department, Virginia Tech), with over 800 (pers. comm.), Carl Djerassi (chemistry) with over 1000, and Linus Pauling (chemistry) with over 700 publications, respectively. Similar results obtain from examination of famous inventors, engineers, and software designers, such as the inventor David Rabinow with over 225 patents and Marvin Camras with over 500 (Petroski, 1992). Qualitative correspondence between this model and actual experience can also be observed in the nonlinear increase in failure probability for software projects of increasing length (Gibbs, 1994).

The range of productivities predicted by this model for more typical full-time research scientists with average skill levels ($p = 0.80$) ranges from 24 for very difficult problems to 175 for very simple problems. Numbers between these extremes are quite reasonable for actual full-time research scientists at the lower range of productivity. Teaching responsibilities of course diminish these numbers considerably. Thus, the p values, the structure of the model, and the number of steps simulated lead to quite reasonable results in comparison with real scientists, and may safely be extended to other professionals on the basis of data from engineers, inventors, designers, and software producers.

In Figure 2.2, relative productivity is shown for short and long productions. That is, the value plotted at a given value of p, $\mathrm{Prod}_{1.0}/\mathrm{Prod}_p$, allows comparison between each level of p and the most

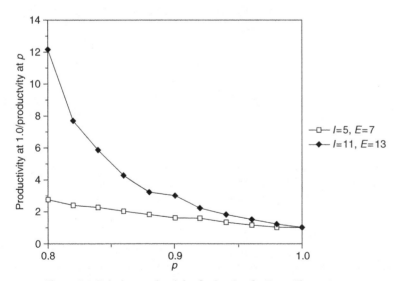

Figure 2.2 Relative productivity for levels of p. From Figure 2.1

productive. Relative productivities are relevant here because professional advancement opportunities are based on relative performance measures. The difference between these two curves indicates that the relative advantage of choosing simple, few-step problems (or those that are more parallel or iterative) increases as p decreases. For few-step problems (the lower curve) the most skilled professional ($p = 1.0$) is 2.8 times as productive as one at $p = 0.8$. The relationship is rather linear with shallow slope, suggesting that to a major extent it is possible to substitute perspiration for inspiration on this type of problem.

Thomas Edison, famous for his "science is 99% perspiration and 1% inspiration" quote, actually followed this strategy. Much of his work was trial and error or incremental (e.g. he tested hundreds of items for his light bulb filament), and he put in record-breaking hours in his laboratory. He also had an army of assistants. On projects that involve few steps, therefore, sufficient variation in the number of hours individuals work or the number of assistants they have can account for the large variation in individual productivities. The upper curve in Figure 2.2 represents the opposite case. For long, multistep problems such as difficult proofs in mathematics, the effect of p is very nonlinear. In the upper curve, the person with $p = 1.0$ would be 12 times as productive as one with $p = 0.8$. Perspiration can not be substituted for inspiration on such problems because there are insufficient hours in a day to make a difference. In fact, if long hours lead to even a slight decrease in p due to fatigue, then Edison's approach will be counterproductive. In addition, an army of assistants may not necessarily be helpful (and may be distracting) but work habits and various personality factors such as persistence might alter p slightly and have a major effect on long multistep problems because of the strong nonlinearity of the relationship.

Discussion

This model suggests a reexamination of our focus on idea generation as the alpha and omega of creativity. Our typical picture of the highly creative person is the individual bursting with breakthrough concepts and ideas; the "Eureka" experience is common for such a person. Most books on creativity focus almost entirely on this step of the creative process. Whether this flood of ideas is translated into a flood of finished products, however, depends on the length and difficulty of the elaboration, execution, and communication stages. Thus, starting with the basic model presented above, it is useful to consider how various factors affect p, the nature of the problem (iterative versus sequential), and the number of steps at each stage.

Consider Michael Faraday. Because of the nature of his problems and his skill as an experimenter, he could perform dozens of experiments per week. Thus, even though he had many hunches that did not pan out, he still could achieve a high productivity (measured as successes). In contrast, in some fields progress becomes rate limited at the execution stage. For example, in field ecology single experiments are very time consuming, and the logistics of running them dictate that a single researcher can only handle a few such projects at a time. Observing lion behavior in the field may require thousands of hours. One only needs a few good ideas per year in this context, and 50 do not help at all. Rather, meticulous attention to detail and a long attention span are essential. Clinical trials can also be very time consuming and are a major rate-limiting step in medical research. The communication stage can also be rate limiting. Richard Feynman hated to write. Much of his work saw the light of day only because of his coauthors (Feynman, 1984).

Creativity can also have an effect on productivity. Creativity leads some individuals to recognize problems that others do not see, but which may be very difficult. Charles Darwin's approach to the speciation problem (F. Darwin, 1958) is a good example of this; he chose a very difficult and tangled problem, speciation, which led him into a long period of data collection and rumination. This choice of problem did not allow for a quick attack or a simple experiment. In such cases creativity may actually decrease productivity (as measured by publication counts) because effort is focused on difficult problems. For others, whose creativity is more focused on methods and technique, creativity may lead to solutions that drastically reduce the work necessary to solve a problem. We can see an example in the development of the polymerase chain reaction which amplifies small pieces of DNA. This type of creativity might reduce the number of steps or substitute steps that are less likely to fail, thus increasing productivity.

The above factors, I believe, explain why little relationship has been found between creativity and productivity. Creativity can have a large positive effect, a negative effect, or no effect, depending on the stage of the problem solving process to which it is applied and the nature of the bottlenecks inherent to the specific field. Further, the role of work habits versus skill level will differ on few-step compared with multistep productions. Taking creativity into account in the context of the whole process of creating technical productions should allow for a more complete understanding of its role.

This model also explains how those who are not brilliant can still do good work. If p is lower, it is still possible to finish difficult projects if the nature of the problem allows indefinite repetition of any step until it succeeds (i.e. the problem allows an iterative approach). This iterative strategy leads to lower productivity, because each stage takes longer, but increases quality. Some writers, for example, must edit a document through many drafts but achieve good quality in the end. Peer review also raises p and is most effective if it occurs not just at the final stage (when it may be too late) but at the early stages as well. Repeated review during a project decreases the nonlinearity inherent in the process by reducing the number of steps that must be repeated when a failure is detected. Such reviews can be particularly valuable for software projects. Peer review is also analogous to redundancy in circuit design: independent reviewers are very unlikely to make the same mistakes as the author. Thus, peer review (of the constructive variety) has a power-function positive effect on quality. Unfortunately, pressures on the individual scientist today make it difficult for peers to find sufficient time to review the work of others.

From the perspective of this analysis, it can be seen that standard views of the "genius" are seriously mistaken. When there is too much focus on the flash of insight type of discovery, and simultaneously the genius is held up as the exemplar of creativity, then the ordinary person feels hopeless to ever do quality work. Who can expect to produce a great theory in a single flash? The reality is far different. Most historical figures of exceptional renown in fact applied an iterative problem solving approach. They did not just burst forth with finished products in an unpremeditated way, but sketched, revised, reworked, and polished. Darwin spent 20 years working on his theory of evolution. Throughout this period he struggled with the parts that did not seem to fit or that were weak. Picasso tried out huge numbers of ideas in rough works before settling on a finished product. The same can be said of most such figures. What perhaps distinguishes such people is not the brilliance of their initial insight, but the fact that they keep working on a problem until the result is exceptional. They can recognize when a solution (or work of art) is incomplete or inadequate in some way, and keep trying to improve it. Many people conceived of the light bulb, but only Edison and a few others kept at it until they had one that worked. Interestingly, Edison also realized that he would never sell many light bulbs unless an electrical power network existed, and he therefore established one. He thus followed through on the implications of his invention more thoroughly than others did.

This multistep view of productivity has implications for improving the productivity of the individual scientist. Writing skills are highly amenable to training and practice. Particular attention should be paid to bottlenecks in the project that are time consuming or have a high risk of failure. Such bottlenecks can be overcome, for example, by using meta-analysis to combine information from a series of clinical trials or studies (Mann, 1990). This approach is very useful because conducting the "perfect" single experiment is highly unlikely in fields such as medicine. A bottleneck at the ideation stage may be overcome by certain work habits (Loehle, 1990 and previous sections in this book). Development of new experimental techniques or instruments may also eliminate error-prone bottleneck points or make them less time consuming (Crease, 1992; Hall, 1992). Molecular techniques such as PCR, gene sequencing, and hybridization have removed bottlenecks in the study of biology and genetics.

We may also apply this model to the problem of the professional whose productivity is deficient in spite of their skill and industry. Some professionals are very busy and do very good work, but much of their work is never completed in a finished form. A close examination of the files of many scientists will show dozens of manuscripts that have been started or are even almost finished but that will never see the light of day, as well as drawers full of data that will never be analyzed. Inventors end up with file drawers full of partial designs, almost finished prototypes, and partially written patent applications. Writers have partial stories and half-finished novels filling their office. All of these people know that greater productivity would lead to greater rewards, to promotions or raises or an enhanced reputation, so why do they leave so much unfinished? Are they lazy? Observing them in action clearly rules out laziness as a factor. I believe the answer lies in the fact that for any such complex product, there is a high chance that at least one step will prove problematic (will be a failure). One key piece of data will prove elusive, one design criterion will be left unfulfilled, one analysis will prove too difficult, and the project will be put aside for a while. Since there are always many things to be done that are not so difficult, that are even enjoyable, projects in which a step has failed may be put on the back shelf and left and eventually they become forgotten or outdated. In most such cases, the failed step can be repeated or fixed, but the level of effort for that step may be greater than the average required for other projects. The single equation holding up a manuscript might require days of sustained effort of a rather unpleasant sort, or a key experiment might require equipment that is not available in one's own lab. Thus because these single steps appear to be too formidable, the entire time spent on that project may go to waste. All of this results from the inevitable probabilities of multistep problem solving. The productive professional, the strategic thinker, recognizes that key tasks may require more effort than average, and also recognizes that such extra effort on certain tasks is warranted to avoid losing the time invested on a project prior to the encounter with the difficulty. These difficult or problematic steps may also require teaming with experts in other domains.

This view of productivity has implications for the management of professionals and of technical projects. In hiring, subsequent productivity apparently can not be predicted from test scores or college grades (Simonton, 1988). On the other hand, an established record of productivity is probably stable over a lifetime (e.g. Parker, 1989; Simonton, 1988). The model presented here suggests that more task-relevant tests are needed for predicting professional performance. For example, measures of attention span and task persistence might be more relevant to predicting technical productivity than measures of either creativity or subject knowledge. Creativity tests that focus on analogy, novelty generation, etc. are unlikely to be relevant at all unless the task of the professional is specifically to generate novelty (as in advertising). The model presented here also suggests that interruptions and distractions can be very detrimental to total productivity. This conclusion is supported by historical examples. Many famous scientists have isolated themselves for long periods. Skunk works in industry have led to extraordinary accomplishments because the group was protected from interruptions and because they allowed rapid feedback at each step, so that single steps could be repeated immediately (Peters and Austin, 1985). In contrast, when control is exerted from the top (e.g. in the name of oversight), two things happen in addition to interruptions. First, the number of steps increases greatly. Military procurement is a classic case in which hundreds of people (and steps) are involved in writing specifications for a product. The probability of a mistake rises nonlinearly with the number of steps. Second, the long planning and specification process may cause a prototype not to be built until "the end," when mistakes may require massive reworking. Thus, excessive control and accountability can lead paradoxically to nonlinear increases in time or money expended and chances for failure. The recent high-tech systems fiascoes such as the Sargent York military vehicle in the United States and the Hubble telescope thus may not reflect stupidity but merely a specification-and-design process with too many steps and players and not enough iterative problem solving (see also McDonald, 1990). Specifically, when a task becomes extremely sequential (as are many large software projects), it may become nearly impossible to prevent major failures, however careful the participants. This was recognized as one of the reasons the United States Star Wars (Strategic Defense Initiative) program was unlikely to succeed and why it was finally killed. In this context, we can clearly see one of the benefits of virtual reality. When applied to buildings, a virtual reality system allows a trial of the building as designed. The ability of a person in a wheelchair to navigate the building, for example, can be tested. The safety hazard of projecting pipes and equipment in a factory can be assessed by a "walkthrough." In this way the building can be assessed before it is built so that changes can be made cheaply, at the design stage. This in essence converts a sequential process into an iterative one, with consequent increased end product quality and reduced costs.

Productivity has become a principal measure of success and a key to promotion. The nonlinear model presented here suggests, however, that problems can arise from simple minded publication counts as measures of professional output. Choosing small, simple problems (or salami slicing) can radically increase output, and therefore rewards, out of all proportion to the significance of the work performed. Tackling the really difficult problems may give the greatest payoff for science or the corporation but not for the individual professional if he is judged on patents or number of publications. Promotion practices may thus be inherently unfair when comparisons are made between, say, biochemists (inherently high output rate) versus evolutionary biologists (inherently low output rate). In addition, it is possible to achieve higher productivity by avoiding difficult problems, for example by applying standard methods to well-defined problems. Innovative ideas are both more difficult (more steps, at least at the ideation stage) and riskier (p may be lower if new methods or tools are used). The discoveries that receive the most attention may require several really innovative steps and thus be very unlikely. In the extreme, high productivity can be achieved by using mass production approaches or by doing many descriptive studies (e.g. species lists), as noted with examples by Simonton (1988), in which case the number of steps at stage 1 is greatly reduced (new ideas are not really needed), and p for stage 2 is increased (because well-tested methods are being used). These considerations point out the pitfalls of measuring success purely by publication counts.

There are implications of this model for time management. In graduate school, the student has a single research project, after which they move on to a job or further study. If the scientist takes this approach, the result is much lost time. Let us say he finishes a project and then starts thinking about the next one. Some months may be consumed while he thinks about it, searches out literature, etc. Then, perhaps he needs to write a proposal and submit it for funding. Or, perhaps he needs to order equipment or wait until his next student starts up his program. These delays can easily run into many months or even years of waiting. This is clearly a drag on productivity. Thus the scientist should always be thinking about the next project, searching out literature on it, looking for funding opportunities, and so on. During a project, also, there can be delays. For example, after field samples are collected they may need to be sent off for analysis. Or, at some stage of the work the data may need to be sent to the statistical member of the team who may not send it back right away. Or equipment may break or lab animals die, leading to delays. These types of delays can add up to months, and can easily double the time a project takes. It is thus crucial to have several projects going so that while waiting for something the researcher can work on something else. However, having too many projects going can tax the memory and organizational skills of the scientist or mean that none of them get done in a timely fashion, and are thus no longer state-of-the-art when submitted. The number that is "just right" is something that must be determined for the individual scientist in the context of the type of research he does. For example, if the person is doing field work, it may not be possible logistically to overlap two such projects because they are at remote locations. However, while waiting out rain or other delays in the field the researcher might be able to work on a grant proposal or a book chapter. This simple process of juggling helps account for the high productivity that some scientists realize.

These results also have implications for educational systems. Most educational systems are geared toward adequate performance, but most real-world jobs require far higher levels of performance. If an auto mechanic knows where only 85% of your carburetor parts go, this is not a "B" performance but rather rates an "F." This is particularly so for highly sequential tasks. Unfortunately, most school systems almost completely lack emphasis on the skills or work habits that lead to successful multistep problem solving (Hunter, 1978; Thelan, 1972). Term papers, long-term observation, diaries, science projects, building machinery, etc. are all largely skipped because they are too demanding of the teacher, in favor of multiple-choice tests based on short-term memorization. Thus, parallel rather than sequential or iterative problem-solving skills are emphasized. Subject mastery rather than minimal competence should also be emphasized. Problems in science and business require not multiple choice, but instead multiple-step solutions. Our educational system needs to reflect that reality. It is notable that the U.S. graduate education system, praised as among the best in the world, is strongly hands-on and apprentice based. The graduate student generally becomes a junior partner on a team and is given tasks to do that require original thought, but within the structure of the overall project and with the guidance of the advisor if problems arise. The sink-or-swim approach to graduate education used by some professors is not optimal and only works effectively with particularly mature and independent students. On the other hand, the graduate-student-as-lab-technician approach denies the student the chance to develop critical thinking skills and does not produce an independent scientist. In fact, such graduate students may remain as lab technicians and never mature into scientists. Thus an optimal apprentice program allows independence but provides some guidance and feedback on performance.

Conclusion

The new operational model presented here can tie together cognitive models of problem solving and sociological models of the scientific process. Such factors as available technology, reward structure, and marginality (Simonton, 1988) affect the type of problems chosen (short or long) and the base level of p for the professional culture as a whole (we now routinely solve problems that would have been too difficult for the average professional 100 years ago). For example, members in an elite department or group can increase p by having access to feedback, advice, techniques, equipment, and appropriate collaborators. The model presented provides a quantitative framework for converting test scores (e.g. college grades, standardized test scores) into predictions about productivity that specifically account for the nonlinearities involved. The model also makes specific predictions about how test performance should change on simple versus multistep problems as a function of skill level. The importance of bottlenecks to productivity is pointed out. Differences between fields are also explained. Some fields are largely focused on short problems, whereas others (e.g. economics, ecology, sociology) face difficult-to-control, complex systems that cause studies to be often long and tedious. The model developed is applicable to industrial or technological problem solving and large-scale software projects as well as to individual productivity. Overall, critical path analysis of technical problem solving can provide a new perspective that links cognitive and sociological aspects in a single framework and that is based on observables such as problem type, operator skill, and problem length.

This analysis clearly shows why novelty or creativity alone are inadequate. Creativity only becomes innovation when a finished product is produced. Producing finished products requires attention to the entire sequential problem solving process. The final critical step in innovation and problem solving is the testing of the value or validity of the finished product, discussed next.

2.4   REALITY CHECK

In previous sections, the process of discovery and the harnessing of one's mental machinery in the service of discovery, problem solving and invention were discussed. By definition, when creating something new or solving a novel problem, we are stepping somewhat into the unknown. Clearly, in such cases we can not look up the proper answer in the back of the book. On the other hand, it is not always clear and evident that our discovery or problem solution is correct, useful, or the best. For an artist or composer it is only necessary that the product be pleasing to the author and/or the public. No further test is necessary, in spite of the presumed central role critics give themselves. In other realms of endeavor, however, it is essential that some type of test be used, some reality check, to verify the solution we have arrived at. This is particularly critical for complex systems. There are two types of tests that we may apply. First, we may be somewhat confident of the product of our work if we can be sure that the problem solving process itself did not lead us astray. In this case we can not be sure that our result is correct, but we can assert that we did not make any obvious blunders. This is an internal reality check. Second, it is useful, if possible, to check our results against some objective standard. This is an external reality check. These are discussed next.

Internal reality check

There are several types of pitfalls that lie in wait for the strategic thinker. Our mental machinery is very unreliable in certain domains, and has some distinct blind spots. As Margolis (1987) has discussed, our minds work primarily by recognizing patterns. We are very quick and proficient at recognizing faces, musical tunes, and places. This same capability aids us in making discoveries by finding new patterns, new structures, and new connections between data and in discerning data from noise. However, just because the brain detects a pattern does not automatically mean that the pattern is real. As children, the darkness is full of monsters and the clouds are full of Disney characters. As adults we misread the pattern of stock prices, of economic indicators, and of spousal moods. The pitfall here lies in our tendency to accept these patterns that we observe as real because the process by which we discover them is hidden. Just as we are unaware of how we identify a face or a song but are nevertheless confident about our identification, we are also unaware of how we identify other patterns and thus have no opportunity to question our judgment. Because of this it is crucial that we examine the structure of the patterns we observe, construct, or "discover" out in the open as much as possible. The use of logic and examination of pattern or structure coherence are particularly helpful in this regard. It is also very useful to become aware of how we assess probability (risk, chance) in decision making and in weighting evidence that goes into our judgments.

Faulty generalization

As mentioned elsewhere in this volume, our reasoning processes are far from foolproof and rigorous. One universal flaw is the Law of Faulty Generalization (A. A. Furutan, pers. comm., Haifa, Israel, June 1995). This law states that we tend to overgeneralize from single incidents to entire classes. If we are in a distant city and someone helps us when we are lost, we believe the entire city is full of kind people. If we visit another city and get mugged, we will never go there again. We can understand the basis for this response in ingrained survival mechanisms. For example, it is extremely important in an evolutionary sense that you never forget a certain food that is poisonous and that made you sick when you ate it. The second time you try it you might not survive. Thus people can develop extremely strong food preferences based on experiences as a child that made them gag (even when a hair in the food, and not the food itself, caused the problem) or caused them to get sick. The same is true of almost drowning, almost being caught by a lion, or almost falling off a cliff. We are thus not wired to accurately find the long-term odds of falling off a cliff or to accurately sample the people of a city for friendliness, but rather to entrain into memory any particularly good or adverse situation and to generalize it.

We may note that this is the origin of superstitions; a close coincidence in time between some unusual event such as seeing a black cat, and a subsequent serious adverse occurrence. While certain people are far more likely to overgeneralize than others, and thus to become fearful or superstitious or compulsive, we all do it to some extent. Whereas generalization from single encounters with dangerous animals or foods is adaptive, we also generalize about abstract categories such as ethnic groups or places or social situations.

While it may not matter much that you don't like eggplant because the first time you tried it you gagged, so you won't try it again, in other situations this propensity to overgeneralize has serious consequences. When we see a TV report of a plane crash, we overgeneralize (that could be me) and are so horrified that we become afraid of flying even though it is far safer than driving. In puritanical societies, the proscription against adultery (sex outside marriage is bad) has become overgeneralized (sex is bad). We have now overgeneralized in the other direction: if sex is good, then sex with anyone any time must be good, so that all standards become lost. We conclude that if salt in the diet raises the blood pressure in the short term, then salt must cause permanent high blood pressure in the long term, a conclusion totally without merit.

Let us look at a more extended generalization problem: recycling. The concept that recycling could be a good thing for conserving resources is not without merit. During World War II extensive recycling, especially of metals, was practiced. Recycling practiced by industry can clearly be shown to be a money maker while at the same time helping the environment. From this the generalization is made: recycling is good, therefore residential recycling will also be good. Experience has shown, however, that residential curbside pickup of recyclables has been a money loser almost everywhere it has been tried (Bailey, 1995) because the cost of curbside pickup is just too high and the items are too mixed and low value. In contrast, in the factory tons of an item to be recycled are generated at one discrete point, which makes it cheap to collect. A related example of faulty reasoning is the perception that landfill space is running out. A few local examples of landfills that are full have been generalized to a presumption of a national crisis (Bailey, 1995) but closer examination usually reveals that these local cases result from NIMBY (not in my back yard) lawsuits, not from actual lack of suitable locations. Further, modern landfills are filled not just to 50 ft (about 15 m) but to 100–300 ft in depth and are compacted, which makes them last much longer than old ones. This phantom crisis results from using superficial impressions rather than actual data to make decisions. The Long Island garbage barge Mobro that wandered the seas for two months in 1987 was perceived to be looking for a dump that wasn't full, like Demosthenes with his lamp, but that was not the real story. The barge owner was a novice at garbage disposal, and sent the barge down to North Carolina without making an arrangement for acceptance. This caused the dump owners to ask questions (was there toxic waste in the garbage? what was the rush?) and stall, which meant that the barge was not allowed into the harbor (since it would tie up traffic and dock space) and was therefore sent back to sea (Bailey, 1995). Once this happened, the barge became famous and no other dump would touch it. A false conclusion about why the barge was not accepted (no dump space) was overgeneralized to the conclusion that all of our dumps must be full, especially as the barge continued to wander, causing a feeling of panic. Never mind that no other barges were wandering the seas nor were trucks full of garbage wandering the highways. One could not even show that dumping garbage was particularly expensive as it would be if space were short.

We can demonstrate excessive generalization even in cases where experience is extensive. For example, at annual review time for employee performance, the most recent few weeks have a disproportionate influence on the supervisor's performance appraisal. This is not the time to take a vacation, decide to grow a beard, or fumble a project. Similarly, at any point in a relationship, the emotional response of one's partner is strongly biased by recent events which are heavily overgeneralized. What have you done for me lately?

The general solution to this tendency to make false generalizations is to learn to depend more heavily on numbers (rates, odds, rankings, etc.). We should automatically dismiss the single encounter with the person in a new city, because it can't possibly be representative. We do this when a burger makes us sick; our long experience with them makes us realize that this single burger was not representative. In the case of the performance appraisal, the employee should keep a log of concrete accomplishments during the year and provide this as a basis for review.

This approach actually provides a method used to treat phobias. Phobias, while sometimes having a biological basis, also have a strongly learned (or perhaps we should say mislearned) component. The common and successful treatment of phobias such as fear of heights or of crowds or of public transportation is to gradually expose the person to the stimulus in a safe environment using very small doses. For example, for a person afraid of heights, they will first be asked to step up on a very low platform and back down to see that nothing happens. They can then work up to going up in an office building. The emphasis is to overcome the false generalization with lots of valid data, delivered in small, safe doses. Direct confrontation does not generally work because when a panic reaction becomes habitual it becomes self-fulfilling: the person panics at the mall even though there is no reason to, so their experience is disastrous even though objectively nothing went wrong. The fear of panic itself sets off a panic. Thus the small safe doses must be used to allow the person to experience the situation and not have a panic attack. This technique applies learning to extinguish an inappropriate overgeneralization that all situations like the one that triggered the original panic are just as threatening.

As is the case with the other deficits in our reasoning, only knowledge of how we fail to think clearly can allow us to compensate for these failures and therefore to think more clearly.

Logic

People are notoriously bad in their use of logic. While most can handle simple declarative logic, more complex logical trains of thought, particularly if they include negations or exclusions, prove daunting. For example, the commonly used negative argument "No pain reliever proven more effective" is very convincing to many people, but does not in fact mean that the brand in question is superior. It may only mean that no comparison study has been done, or that you can't tell this brand from the others statistically. Thus the computer as a model for intelligence is completely misleading. We neither reason like an expert system (by using sets of rules), nor like a hierarchical classifier, nor like a FORTRAN computational loop. In fact, our formal reasoning abilities are rather limited. For this reason, we can often observe illogical arguments in spousal "discussions" and in political debates. Margolis (1987) argues that we do not think by a process of sequences of logical deductions, but rather in terms of patterns. Thus if certain items seem related or connected (form a pattern), we will believe that they are related in a causal way. We will consequently reason in an illogical way about them (e.g. explaining why they are related) and never notice that our reasoning is fallacious. In particular, the mental operations we use to explain or communicate the patterns we have observed are equally facile at explaining a real pattern and a fictitious one, according to Margolis. For a conspiracy theorist, all types of data fit together seamlessly, even data that later turn out to be erroneous or merely rumor. This means that we must take special care when attempting to reason logically, and should apply particular techniques that are less subject to this type of pitfall.

Another faulty type of logic concerns reasoning by opposition. We often enough experience exclusive relations (one can not be in two places at once, one can buy either item one or item two with a certain amount of cash, the person is either alive or dead, it is either true or false that a certain event happened) that this approach to reasoning comes to seem reliable. When applied to complex issues, however, such reasoning is inadequate. If we argue that pure communism is bad, it does not follow that its opposite, pure capitalism, is perfect. If one spouse is wrong, this does not mean that the other spouse is right. Black-and-white reasoning is prone to reducing a complex issue to an either-or question, when in reality neither alternative may be the best or a compromise may be the best and not either extreme. Thus extra care must be taken to avoid this type of logic because it seems much more conclusive than it is.

Because we can operate fairly reliably on simple if-then type logical statements (if I buy this item, then I will run out of cash), since such statements correspond to our usual sense and experience of cause and effect, such declarative logic does provide a useful tool for examining our reasoning about patterns. We can use it in two principal ways: to examine the premises underlying our idea and to examine the consequences of our idea.

Underlying any discovery, pattern, invention, or structure is a set of premises or assumptions. Such premises are often much more easily examined than is the overall pattern or theory or concept. The successful introduction of the minivan was based on the correct premise that for families a station wagon was just too cramped but a full-size van was too big, too hard to drive, and too expensive. In other cases, the premises may be suspect. For example, if one assumes that there are few homeless and that most of them are mentally disturbed or substance abusers, then one will support a very different social policy than if one assumes that all homeless are down on their luck and eager to take advantage of an opportunity to better themselves. In many cases the assumptions people make about the homeless are based on no data at all and no personal experience but rather on social theories (world views). Reliable data are desperately needed to characterize the population of concern in this case because otherwise incompatible solutions are offered by groups starting with these very different (usually unspoken) assumptions.

If we examine Darwin's development of the theory of evolution, we see a very successful use of the testing of premises. Darwin believed that the patterns he saw in nature could be explained by natural selection acting on heritable variation. At the time, the genetic code had not been discovered, so he could not examine the mechanisms of inheritance directly. On the other hand, evolution occurs so slowly that he could not observe it in operation either. However, after considerable reflection he realized that he could collect data to evaluate the premises of his theory. One premise was that variation in traits is heritable. He showed this to be true, indirectly, by exhaustive collection of data from plant and animal breeders. A second premise was that organisms are all descended from common ancestors. A challenge to this assumption seemed to be posed by life on distant oceanic islands. If these remote plants and animals could not have traveled there, then they must have been specially created, which would invalidate descent from common ancestors. He tested this premise by showing that seeds of many species can float in salt water for long periods and still germinate and by noting that the animals found on these islands all either can fly or can survive long periods without water (e.g. reptiles) and thus can travel on mats of floating plants for long distances.

Historically, the examination of premises has been critical to finding flaws in ideas. For example, Freudian analysis seems to be a coherent system, and it is good at "explaining" many aspects of behavior. However, as psychology has matured, investigations have failed to find any evidence for the assumptions underlying the Freudian paradigm, such as penis envy, the Oedipal urge, etc. As a result, Freudianism is more popular among the public than among psychologists (Stanovich, 1992).

In contrast, it is typical of those holding irrational beliefs (in folklore, crackpot theories, superstition, perpetual motion machines, etc.) that they resist efforts to examine the premises of their belief system. Freudian psychotherapists and Marxists have never been particularly interested in examining their premises and are both notable for their lack of enthusiasm for experimentation, which is why neither paradigm has progressed and why both are rapidly fading from view.

We may also use if-then logical reasoning to examine our ideas, concepts, or inventions by making causal predictions or extrapolations. Because this type of reasoning matches one of the natural abilities of pattern reasoning and involves only simple conditional logic, people have a natural talent in this area (though it is far from fool-proof). We habitually try to imagine the scenario at a picnic so that we will remember to take everything we need. We try to imagine our old age when planning for retirement. This type of reasoning can be used to derive tests of our ideas or to draw out the implications of them. We can imagine how a consumer might use a new product or try to predict the reactions that might follow a foreign policy action. We can try to see what follows if a particular scientific theory is true. Scale models, mathematical models, and virtual reality simulations are particularly useful for projecting the consequences of our ideas or actions.

It is particularly useful to extrapolate our action or idea to see what it entails. For example, I was considering painting my kitchen because of a large stain above the stove. When I followed through this action in my mind, I got to the edge of the kitchen and realized that the wall is continuous with the living room, dining room, and family room. There is no place to stop. Therefore, I matched the paint and just painted above the stove. We may similarly extrapolate the effect of a policy decision. In the United States education community, it has been decided that bilingual education is a good thing. We can debate the logic behind this policy (in particular, the prescription does not follow from the premises, as discussed below), but here I just want to extrapolate this policy. If bilingual education is a good thing, then we should apply it everywhere. However, some schools have immigrants from dozens of different countries. Do we provide bilingual education for them all? Doing so will totally bankrupt the school district (or the nation) by leading to class sizes of one in some cases. If we only provide bilingual education for Hispanics, then we are either being conde-scending (the others don't need help, only Hispanics are too dumb to learn English), or we are practicing favoritism. Thus extrapolating the situation from the area on the Mexican border (large numbers of Hispanics) to the entire United States (multiple nationalities of immi-grants) shows the policy to be impractical.

Let us take another example. It has been argued that extenuating circumstances should be considered in criminal cases. For example, if a woman murders her husband (in cold blood, not in self defense) because he was constantly beating her, there will be a tendency to view her situation with some sympathy and she might get a reduced sentence. It has similarly been argued, however, that other individuals have been traumatized by parents or by society, and that this likewise excuses their criminal actions. The danger here clearly is that when we extrapolate the reasonable argument for abused wives to other cases, we enter a domain where almost anything can be argued. Husbands have argued that their wives abused them verbally or had affairs and that this justified murder. Gang members have argued that they are hopeless and oppressed and therefore their lifestyle is excusable. There is virtually no end to the excuses that can be made in this fashion, and we end up weighing imponderables about who has been victimized or abused sufficiently to be excused from a crime. Thus again extrapolation shows this policy to be unworkable. We must insist that there are always alternatives to violence or crime, and that while we offer sympathy to the abused or oppressed, we can not excuse criminal actions.

A difficulty with examining premises in practice is that the reasoning processes many people use are not consistent, such that their conclusions often do not follow from their premises. That is, the truth or falsity of their premises (assumptions) may not bear on the truth or falsity of their conclusions. There are several reasons for this.

The first type of reasoning failure is that we are often working with incomplete information, though the incompleteness of our set of premises is not always obvious (except in hindsight, of course). It is a fundamental error to attempt to draw conclusions logically from a set of incomplete premises. Such a situation is indeterminate and many different conclusions can be shown to be compatible with the given premises. Consider a designer creating a consumer product for outdoor use. He considers sun, rain, heat, and dirt as factors affecting product durability, and designs accordingly. However, because he lives in Iowa, he does not envision the widespread use of the product by the ocean, where it corrodes overnight. I have in mind here two brands of folding lawn chairs I bought in South Carolina. One was all aluminum and held up well by the beach, but the other was steel and rusted out in no time. There are only two types of protection against this difficulty: first, actively seek out all possible impinging factors; second, test any decision, invention, or theory against reality (see below). We observe the consequences of failure to guard against this incompleteness pitfall in the lives of grand theorists (social, political, or other) who are so confident of their reasoning ability that they never stop to think that they might be missing something (Johnson, 1988). Thus are born an endless series of pop psychology prescriptions for improving your love life, self-esteem, weight, addiction, etc.: each is based on only a partial set of correct (to be generous) premises.

A related problem is caused by faulty reasoning per se, such that the conclusions do not follow from the stated premises. We observe this frequently in politics. One cause of this is the tendency to jump from premises to prescription. For example, from the premise that crime is high, we are immediately given the prescription that prison terms must be longer. The analysis is presented as though the prescription follows inevitably from the premises, which it does not. The prescription is only one of many possible solutions. An alternate prescription might be to intensify police presence and to make jail time more certain (rather than longer) for an arrested criminal. This would make capture and punishment seem more likely to the criminal. A second cause of conclusions that do not follow is the single causal factor fallacy (Stanovich, 1992). For example, from the fact that childhood sexual abuse is traumatic and can cause adult adjustment problems, certain therapists have drawn the conclusion that every adult who has problems (or even every adult) is the victim of sexual abuse. This conclusion does not follow, however, because there are many reasons why an adult can have problems besides sexual abuse.

A further difficulty is that in many cases the reasons given to explain a decision actually have nothing to do with the decision. Often the articulated "reasons" are nothing more than rationalizations or justifications. People are quite good at offering "explanations" for chance or even nonsensical events (Margolis, 1987) and are even better at it when trying to justify their actions. The "reasons" offered for the choice of a particular job candidate often obscure irrational decisions or outright biases. The steel company president who goes wild and buys an oil company while ignoring the modernization of his mills may offer elaborate logic for his decision based on business cycles and return on investment, but his real reasons may not be rational at all (maybe the oil business is more glamorous to him than the steel business). One should likewise be suspicious of the reasons offered by people for their choices of cars, careers, and houses.

A final difficulty arises from the decision criteria people use when reasoning from their assumptions. There are usually several conflicting goals in any decision making context. If the wrong one is chosen, the resulting decision or product will be deficient. For example, let us say that a company sells a low end automobile. Some complaints come in about various missing features. The goal of fixing these "problems" conflicts with the goal of keeping the price down. If lots of fancy features are added to a slow, small, boxy car, it will no longer be the cheapest car, but it will not attract higher end buyers. Better gas mileage on a luxury car may not expand the market for that car; this is not a major criterion for such buyers. Girls may choose a horrible spouse because the criteria they use for picking someone to date (wild, fun, a heavy drinker, a flirt, unpredictable) leads them to only become acquainted with, and therefore to get married to, a real bum. A common response to conflicting goals is to look for a compromise, but often an exclusive choice is necessary, as in the auto example above, because a compromise actually satisfies no one. One can either become a doctor or an artist, but very few can do both. In the dating context, conflicting goals lead to a cycle of dating and breaking up, because the fun person chosen for dating becomes annoying or impossible to tolerate in a long-term relationship. Thus the problem may lie in the decision criteria and not just in the premises or methods of reasoning.

Thus we can see that it is necessary and useful to examine the premises behind a decision, design, or theory, and the reasoning used to get there. Everyone believes they reason things out like Sherlock Holmes, Matlock, Colombo, and Agatha Christie, but few of us really can or do.

Coherence

As mentioned previously, the tendency to detect pattern is strong enough that even disconnected events or forces can seem to be part of a pattern. We see this tendency run amok in the paranoid schizophrenic's tendency to weave random, unconnected events into a vast government conspiracy. We can counter this tendency by checking for coherence. In a coherent series of events there is some type of logical or causal dependence. A check for coherence is also a search for pattern, in a sense, but is a search for items that make a pattern by being causally related. Following a hurricane, for example, we may observe high prices, emotional distress, and looting. All of these are caused by the effects of the hurricane and are thus coherent (though independent) effects. A corporate reorganization can affect productivity, morale, and stock prices, again in a causal and coherent way. In an incoherent "pattern" there is no such connection. I recently came across a hollow earth enthusiast's letter to a major government research lab, seeking their support for his theory. He attributed auroras to light escaping from a sun inside the earth via a hole at the North Pole, UFOs to flights by the people living inside the Earth, etc. Not only were the "facts" he put forth completely incoherent with respect to each other (there was no necessary relation between them), but they were completely out of sync with dozens of other facts (a hollow earth would collapse and the oceans would run into a hole at the North Pole, for instance).

As another incoherent system we can look to New Age philosophy, which incorporates astral projection, past lives, healing crystals, vegetarianism, UFOs, and New Age music. There is no necessary relation between these separate beliefs. They do not derive from a set of premises, and are not logically or causally related. In fact it is possible to substitute alternative beliefs for any one of the ones listed in an arbitrary way. It is thus a collection, not a system.

As an example of an incoherent scientific theory we can again turn to Freudian psychoanalysis. As Stanovich (1992) notes, the component parts of this theory contradict one another. Consequently one can come to almost any conclusion that seems convenient. This is a telltale sign of an incoherent system: it is too loose and permits too much, so that anything goes. Stanovich cites as another example folk wisdom. There is an aphorism for every situation, but they can be contradictory ("look before you leap" but "he who hesitates is lost").

Incoherent beliefs or systems are held by people with magical thinking, by children, and by the mentally ill. Most people are guilty of incoherence in at least some part of their lives, for instance in their views on love or politics. We may believe that government meddles too much, but at the same time expect it to fix every problem and right every wrong. In love, a girl's stated views on what makes a good husband may conflict sharply with the type of guys she dates. A guy may sleep around but still expect to marry a virgin. In contrast, in science and engineering there are multiple logical and causal connections between different components of the enterprise, each contributing to and supporting the other, and giving high coherence. Thus coherence is a very useful tool for examining our beliefs.

Figuring the odds

In strategic problem solving, one is often faced with risks or with chance events. In planning a factory expansion, one must factor in the odds that interest rates will rise. One must balance extent of insurance coverage with the risk that it will be needed. Unfortunately, people are extremely bad at estimating odds and risks. In cards and games of chance, people seem to be almost without a clue as to their odds, which makes casino owners very rich. Slot machines are designed to make a noisy racket as the coins fall into the tray, thereby creating the appearance of many winners and big winnings. In estimating risk, people are equally bad. People feel perfectly safe on a bicycle, but bicycles are far from safe. Many parents are terrified that their child will be kidnapped and murdered, though they are actually 100 times more likely to die in a car crash (Stanovich, 1992). Dramatic events, such as plane crashes, seem much more prevalent than they are. Similarly, people misjudge the risk of stroke because death by stroke is much less dramatic than is death from cancer or heart disease. People's behaviors are often influenced by these incorrect perceptions.

Strategic problems often involve a probabilistic or risk component, and failure to recognize that people are quite bad at assessing risk or probability often results in adverse outcomes. A mother in a quiet suburban neighborhood may refuse to let her children ever play outside for fear of kidnapping. A manager may fear making a mistake so much that he refuses to make a decision, not realizing that failure to act also carries risks. In certain hurricane-prone areas such as Florida, insurance companies have long had actuarial tables showing the risk from storms, but during recent decades they ignored their own perfectly sound analyses in favor of the perception that risk was low because there had been no major storms for a while. Thus during Hurricane Andrew in 1993, insurers were exposed to excessive liability, especially from ocean-side homes and cheaply made homes. In fact, the course that insurance companies take, figuring out the odds of various risks explicitly, is the proper course and it usually serves them well. Their failure in Florida was because they ignored their own calculations.

This actuarial type approach is the only protection against the failure of our mental machinery to deal properly with probabilities and risks. In the context of any strategic problem, whether product design, urban planning, invention, etc., one must specifically and quantitatively account for chance events. In engineering design, for example, failure analysis is a standard, quantitative design criterion. In military campaigns it is standard to hold troops in reserve in case of unexpected reversals. In science one is expected to account for the odds that one's experimental results were actually the result of chance.

Thus we can see that there are a number of tools we can apply to examine our ideas, models, designs, and theories to see if they pass muster in terms of not being the result of obvious mistakes caused by the imperfections in our reasoning abilities. Unfortunately, it is far easier to apply these techniques to the examination of the work of others than to ourselves, and thus more objective, external reality checking is also necessary.

External reality check

Given that one has taken precautions against the pitfalls noted above by checking for coherence, examining premises, and making any assessments of risk or probability explicit, it is still not guaranteed that one's result is correct or useful. A logical and coherent theory can be wrong and a clever invention can be too expensive to sell or never work properly. For this reason, an external reality check is necessary. Three basic types of external reality check are available: breaking it, peer review, and experiment or testing.

Break it

Once you have created something, it is very important to break it. Of course, the very last thing that the creator of something novel has the heart to do is break it, but it must be done, for if the creator won't break it the public will. If the creation is a building design, it will be found to have some type of flaw by someone and many will think it is ugly. If it is a piece of software, you can be certain that a certain number of users will make it fail. The perfect assembly line is subject to a change in the products that it will be expected to produce. Even the best scientific theory has some kind of flaw. If the developer of a product can summon the courage to break it, he is in the best position to envision the possible ways to do so, and perhaps to thereby fix it. There are several specific topics in this respect that can help us focus our thinking about the breakage/failure question.

Physical breakage

For most physically realized creative products (i.e. not a theory or abstraction), it is useful to search out the possible modes of physical breakage. Every parent with small children has had to (or tried to) prevent them from inserting various objects into the video recorder (VCR), but one must wonder about whether the VCR designers envisioned this problem. In the same vein, cassette tapes don't melt in the car, but video tapes do. If you wish to return a tape on the way home from work, you can't leave the tape in the car but must risk embarrassment if you take the kids cartoons or your favorite horror movie to work in your briefcase. Thus modes of breakage may turn out to be common even with widely used products.

Let us consider "breakage" for a building. An obvious type of breakage is earthquake damage. Careful consideration of how an earthquake operates (e.g. maximum lateral velocities and displacements, vibration frequencies) is necessary to design against this type of damage. An unusual type of "breakage" occurred in 1992 in Chicago when the Chicago River broke into a system of old freight tunnels that connected to the basements of major office buildings. The water shorted out most power to the entire downtown Loop area and proved to be very difficult to get rid of. Many large buildings had their heating and electrical stations on these lower levels, and their owners had never really considered flooding as a risk. The cost of this disaster was in the billions of dollars.

In general, it is safe to assume that if some mode of physical breakage exists, then it will happen eventually. If that mode involves human misuse, then it will probably happen on a regular basis. Any household item should be able to survive being dropped, for example. If it shouldn't get wet, it probably will. Such abuse should be designed for.

How do you break a theory? One way is by considering limiting cases. What happens at very high speeds or temperatures or pressures? Would my theory still work? This is how Einstein demonstrated the flaws in Newtonian theory. In ecology, asking what would happen as animals become very rare led to the identification of the Allee effect: that animals might have trouble finding mates at low density. In economics, we might ask, what if economic agents do *not* have perfect information or are illogical? And it turns out that theories based on omniscient utility maximizers fail if you relax those assumptions.

Side effects

It is well known that almost every medicine has some sort of side effect. We usually watch for them. We are not so alert to side effects in other contexts, but they are just as inevitable. That is, we should actively and explicitly search out side effects.

Let us consider a high rise building again. What could we mean by "side effects" for a building? As many pedestrians chasing their hats down the streets of the "windy city" Chicago can attest, high rises can funnel and amplify the wind, creating almost gale force gusts during storms. Under such conditions I have seen an entire trash can full of inside-out umbrellas. A second side effect, noted particularly in water-front locations, is that a large building can block the view for other buildings, thereby lowering their value. A skyscraper also typically displaces small shops, thereby making a neighborhood less livable. Thus we can see that side effects are not so hard to find, even for a building.

In ecology, side effects are pervasive. A major impetus to the environmental movement was the documentation of the side effects of industrial activities. Many human activities create environmental side effects which may even occur miles away (as with acid rain). Tracing and managing these side effects remains one of the major tasks of ecology as a science and as a management art.

Identifying side effects is an effective technique for guarding against problems that otherwise seem to come out of the blue.

Difficulty in use

No product will be universally easy to use. This is obviously true for consumer products such as appliances, tools, and bottle caps, but applies elsewhere as well. A revolving door seems like a simple thing, easy to use, until a parent tries to go through it with a stroller. Computers have been notorious for the difficulties many people experience with their use. The increased ease of use personal computers afford has contributed to their growth versus mainframes.

We may equally apply this question to abstractions. When a new scientific theory is introduced, we should anticipate that someone will have trouble grasping (using) it. In the case of evolution, this was largely the public and the religious community. For other theories it may be other scientists who have trouble adjusting to a new world view.

It is the frequent experience of inventors that the public "doesn't get it," of companies that ordinary products lead to many complaints about them being hard to use, and of scientists that their work is not understood. These all result from the fact that consumers will not find a product or theory as easy to use or grasp as the inventor does. This should be kept in mind.

Your worst nightmare

A good way to scare up some of the flaws in a product, idea, or design is to think about what would be your (the inventor's) worst nightmare about it. For a toy, this might be for a child to get hurt. A scientist might have a nagging fear at the back of his consciousness that something might be wrong with his new theory or that an experiment might be flawed. For an inventor, his worst nightmare might be that his idea has already been patented, and that he is therefore wasting his time.

The reason a Worst Nightmare is so useful is that in thinking about and working on a product or theory, thoughts about possible flaws have usually occurred at least in passing. Often, really serious flaws may occur to the inventor, but there is a strong tendency to suppress these thoughts "until later" since it is in truth unreasonable to expect an early stage design or a prototype to meet all possible objections. Once these difficulties are pushed aside, it is too easy to keep them out of the picture, but if really serious they may continue to resurface as nagging doubts or fears. It is best to face them directly once a project has achieved sufficient proof of principle. At this stage bringing the feared flaws out in the open can provide a critical reality check and either improve the product or allow a timely retreat from it if the flaw is real.

Change of environment

Another way to "break" a product is to change the environment in which it is being used. For the developer, this means thinking about the consequences of a change in environment. The gas-guzzling cars of the 1960s, for example, were damaged by the change of the gas-price environment in the mid 1970s. The typewriter has declined due to the introduction of the computer. In fashion, it is well known that a change in public taste can lead to the "death" of a product line. In fact, clothing manufacturers actually plan for such changes in environment. Many products have also failed when subjected to a new physical environment, from standard motor oils that freeze up in the Arctic to house designs that are inappropriate for the tropics.

We may also consider a change of environment in the case of abstract products. Software is always subject to a change of environment. As operating system architecture changes to provide new capabilities and to keep up with hardware advances, the software running on it must also change lest it fall behind and eventually fail to operate in the new environment. Conceiving of a change in environment can also help point out where a scientific theory may fail. In the microworld of atoms, it is observed that the laws of our macroscopic world fail. The lawful economic relations we observe in a developed economy fail in an environment without a monetary system, such as in a jungle village. For example, if there is no way to store wealth (i.e. no banking system), it may make more sense when one has encountered a windfall to give it away in a show of generosity, thereby "storing" goodwill and prestige for the future.

Overall, searching out how we can break our product or theory is a good way to make it stronger. When this effort is not made, the inevitable side effects, customer complaints, lawsuits, and critical reviews come as a surprise. Facing the ways in which a product can be broken requires a nontrivial amount of detachment, discussed in a following section.

Peer review

Every time you ask your friends what they think of the car you are thinking of buying, you are practicing peer review. Reviewers can be extremely valuable because they bring different perspectives to bear on the problem at hand, and they may be less emotionally involved and thus may be more objective. In medicine it is useful to get a second opinion before surgery. In science, every grant proposal and every publication must pass muster with at least two or three reviewers. Movies and plays must endure critics (though often after they are in final form). Some movies are tested in limited release, and modified if reaction is negative. A test audience is only used to predict whether a movie will be popular, not whether it is correct, so in this case we only worry about whether the audience is representative of what the movie will encounter after release. For other types of reviews, it is also possible that the reviewer can be wrong.

The ideal reviewer knows enough about the topic to be useful, is objective, helpful, thorough, and polite. If anyone has seen this mythical creature, please send me his address because I could sure use his help. In the real world, each reviewer comes with various shortcomings and these must be taken into account by the reviewee.

The most serious problem with reviewers is that they tend to be too negative. It is easy to criticize anything, especially if the person views their role as being a gatekeeper. After all, what kind of a critic are you if you can't find anything wrong? New inventions are particularly easy to criticize because they are usually not in final, bug-free form and do not necessarily have a ready market. The original copy machine was rejected by many companies because they viewed carbon paper as sufficient and because the initial copy quality was poor and they couldn't imagine that quality could be improved. The initial zipper design had many failings (Petroski, 1992) such as a tendency to jam open. It took 20 years to perfect the design, create machines that could reliably make them with minimal waste, persuade manufacturers to incorporate them into their products, and persuade consumers to buy them. Throughout this process, it would have been correct to state that the product was deficient, but such a judgment would have been premature. In fact, any new idea or invention must go through a period of maturation and development before it can withstand rigorous scrutiny (as noted previously in this book). A movie concept is not a finished movie. A tentative scientific hypothesis is not yet ready for experimental testing and it might not even be clear how it *can* be tested. No one would expect the outline for a novel to be a best seller. Yet very often a new management scheme in business or an idea for a new product is subject to intense scrutiny and review by management at an early stage of development, and often only an abbreviated, bulletized form of the idea is presented. Under these conditions, it is not surprising that most new ideas are killed off prematurely and without an adequate hearing. The number of inventions that were difficult to market but are now major products is truly amazing. Thus review of an idea, plan, or product is not without a significant risk. It is wise to shield it from review by those in a position to kill it off until the idea is sufficiently developed that it can withstand scrutiny. During this phase, reviews should be sought from those outside the lines of authority. A new idea should not be presented in an abbreviated format because the audience will be almost certain to "not get it." Sufficient detail should be presented so that the concept is conveyed clearly. If further development is needed, the path forward should be clearly stated.

As a reviewee, one must develop a thick skin. Try to glean the useful suggestions out of the negative comments while letting the disparaging remarks about your mental stability and IQ slide. It is often the case that a reviewer has his own pet theory, invention, or point of view that colors his opinion of what he is reviewing. This also must be taken into account.

A second difficulty is that reviewers rarely put the time into their review that it requires. The inventor spends a year on a new product but gets only 30 minutes in front of management to plug it. A scientist spends four months on a project but a reviewer may spend only an hour or so reading the manuscript. Something really novel is difficult to digest, especially upon quick examination. A reviewer with only a partial grasp of the concept is likely to dismiss the idea or focus on trivial or irrelevant aspects of the paper. It is difficult to overcome this tendency. Because most people are busy and since one is usually asking a favor in obtaining a review, one can not control how much time a reviewer spends looking over your work. One can partially overcome this problem by obtaining several reviews and taking into account that each is going to be incomplete. A more satisfactory solution is to pay a consultant or some other party to perform a review. If one is in a position to control budgets sufficiently to do this, it is well worth it, because consultants or paid reviewers will take the time to be thorough, will be more objective, and are not in a position to steal the idea (by contract) or kill the project. One must be aware of the danger that paid reviewers may try too hard to please and may be reluctant to be too critical lest they lose any future work.

The nit picker is a particularly useless type of reviewer, and can be deadly if in a position of power. Nit pickers are often observed in bureaucracies, where the proper filling out of forms and the formatting of reports can come to assume greater importance than the work itself. The nit picker does not merely find grammatical and formatting errors, but focuses on them exclusively, sending a report back to the author with a sneer because of the errors. To such a person the trivial errors invalidate the whole project or report. One can easily identify the nit picker because they are unable to make any useful comments on the overall product, just on the format. A nit picker is useful if they are performing merely an editorial task, as long as they know what they are talking about. Otherwise, they can interfere with real work and can be a deadly bottleneck in an organization by refusing to ever approve anything.

Testing and experiment

Generally, when people think of tests and experiments they think of the procedures of scientists as a model. This is a partially useful viewpoint, but it tends to magnify the difficulty of the process so that people think testing involves expensive laboratories and a Ph.D. This is not the case. Many types of testing are a part of regular business practice and may even be quite simple. Tests are often used such as opinion polls, market surveys, engineering failure analyses, prototypes, dry runs, scale models, project flow charts, and statistical quality control. What all of these types of tests have in common is that they provide objective information about the workability or quality of our wonderful new idea or product. These are next explored in turn.

Surveys

Surveys and opinion polls are a common component of marketing and political decision making. Many new products are test marketed and surveys look for the public pulse on political issues. For test marketing, the issue is largely whether the test audience is representative. Those individuals who are easy to persuade to participate in a market test may not be representative, however. For example, those willing to see a sneak preview of a new movie, which is in reality a test screening, are not likely to be representative of the many people who depend on newspaper reviews and word of mouth before choosing a movie to see. Market niche groups such as urban singles or business travelers may be very difficult to locate and persuade to participate in market surveys. Market surveys do not enable one to evaluate such factors as long-term satisfaction with a product, word of mouth, and repeat buying.

Opinion polls are a whole different kettle of fish. Polls are routinely presented in the media on every issue imaginable, from political issues to how people feel about dogs and spouses. A standard part of these presentations is the measure of error (e.g. plus or minus x percentage points). This is commonly taken as a measure of accuracy, but is in fact only a measure of reproducibility. That is, it measures how likely a repeat of the same survey is to yield the same results. For example, a survey of the readership of a particular women's magazine will have what appears to be high accuracy because of the large number of people responding, but when the results are extrapolated to all women, or even all the magazine's readers including those who tend not to fill out surveys, they may not be representative at all. A second problem is that the phrasing of questions can have a significant impact on how people answer. Although this is well known in sociology, it is often ignored in political polls. For example, to the question "Is health care too expensive" a large majority will answer yes, but to the question "Do you approve of a government bureaucrat deciding whether you can have bypass surgery" a large majority would answer no. Such questionnaires are often phrased in such a way that the sponsoring organization tends to obtain answers that help it. A great deal of cleverness is necessary to obtain answers that are not biased, and private opinion poll organizations do not necessarily take the time to do so. Academic surveys are more likely to be objective in this sense.

A second issue is that people often lie on surveys, even when they know their identity will be kept secret. This is a common problem on surveys of sexual behaviors, but also occurs in other contexts. For example, in a recent survey in Chicago, whites expressed fewer negative stereotypes about various minorities than did members of various minorities about each other or even about themselves. In particular, blacks were more likely to express negative attitudes about blacks than whites were. It may be that this accurately captures the feelings of the two groups, but it could also be that whites are acutely aware that they should not admit their negative feelings, and thus lied on the surveys.

A final problem with surveys as a reality check is that what people say they want and what they do often do not coincide. For example, it is often observed that people will say they are for the environment and schools, but will not necessarily approve a recycling center or a school bond issue. This is because at the point where attitude hits the pocketbook, various goals come into explicit conflict. Thus overall it is necessary to take survey results with a very large grain of salt.

Prototypes

A very useful type of reality check is the use of prototypes. Prototypes have the advantage that they can be touched and examined, and may reveal many deficiencies that are not evident in a verbal description of a proposed object. This is particularly so because each person tends to interpret or visualize a verbal description in a personal way. Even a drawing of a proposed product may be more glamorous or attractive than any physical realization could ever be. Artistic sketches of new or proposed cars are very glamorous but have visual properties that no real automobile can have. A prototype removes this subjectivity. Prototypes have long been used in engineering, where models of a new aircraft, for example, are tested in a wind tunnel. In architecture, scale models are very useful prototypes because a drawing of a new building can easily be made attractive with good artistry, whereas a scale model looks more like the real building. The construction of a 3-D prototype in the computer allows explicit calculation of turbulence in the wake of a plane and how the wings will generate lift. Light penetration into a proposed building can be explicitly viewed. Such prototypes are extremely useful. Scale models of harbors, of highway systems, etc. are likewise very valuable. Prototypes of new consumer products are often produced and tested before full-scale production begins. With prototypes one should always be aware of the scale-up problem: a 600 ft tall building made of steel and glass will look inherently different from a plastic model; the mass-produced version of a radio will look different from the prototype; and the mass-produced cookie will taste different from the sample cookie.

In social engineering, one of the problems is that prototypes are difficult to construct. Almost any social system (welfare, rapid transit, socialized medicine) can be made to sound good (or disastrous) on paper, but without a prototype it is difficult to tell what will really happen. Depending on public opinion for making decisions in this context is a serious problem, not only because of the problems with opinion polls described above, but because without a prototype the public is responding not to the actual consequences of the policies they are being asked about, but to the promises about those policies. For example, one can make any promise one wants about a health care system, but it is more useful to look at particular countries or states with different systems to see what has actually happened, how much it actually costs, etc. Thus for social systems prototypes are a valuable source of information, but in this context the prototypes are provided by different social bodies that have implemented different policies.

Quality testing

Quality testing has become an essential process in manufacturing. For example, a key element in the invasion of Japanese cars into the United States in the 1980s was quality deficiencies in American cars. Clearly, automobile quality had to be designed in, but it also had to be tested for. Tests of product quality are now common and are an essential reality check for manufacturers. Quality testing, if done right, is a vast improvement over waiting for consumers to complain. Care must be taken that one knows what to measure, however, and how to interpret the results.

A type of testing that often seems to be overlooked is consumer operability. It seems to be increasingly common for a product that requires assembly to come with an instruction manual consisting almost entirely of drawings with helpful instructions like "attach handle." In many cases one can not believe that the ability of anyone to follow these instructions was ever tested, and in some cases it is not even possible to do what is depicted in the diagram. Whereas I have purchased a broom that had instructions on how to sweep (!), it is more likely that instructions on consumer products will be inadequate. Testing of these products on naive consumers could help eliminate such problems.

Consumer product testing is a particular type of testing that is done by third parties for various purposes such as safety and product evaluation reporting. In these tests it is common for a standardized treatment to be applied to a product to see how it holds up or performs. The standardization is a key element that removes the subjectivity inherent in merely asking people how they like a product (which is a measure of consumer satisfaction). Consumer satisfaction can be influenced by clever ad campaigns, for example, and may not be related to actual product performance. Just as with opinion polls, however, standardization increases repeatability but not necessarily accuracy. For example, a measure of durability of a dish washing sponge obtained from a mechanical dish washing device may be irrelevant if sponges are disposed of because they begin to smell long before they ever fall apart.

Analysis

Various types of analysis are used in different settings to provide a reality check based on calculations of various sorts. In construction planning, one lays out a schedule to make sure that materials are on hand when needed and that different crews don't get in each other's way. In financial planning, explicit calculations of costs over time are used to test the feasibility of a given project. A dry run of a medical operation, film scene, or military exercise is a test of the verbal description of the operation to make sure it works. Dry runs are particularly important for testing emergency plans (e.g. a hospital evacuation) because when they are used it will be a real emergency with serious consequences if a plan is faulty.

Experiments

For certain types of tests, true experiments in the scientific sense are necessary. It is easy to believe, based on initial results, that a new drug is wonderful and will have a major impact on disease. To get it approved, however, hard data on efficacy and side effects are needed. This necessitates controlled experiments. Conducting and interpreting such experiments was discussed in more detail in Section 2.2 "Discovery as a process," but it is useful to just point out here that even controlled experiments can be subject to error, bias, and misinterpretation. For example, in the world of medical research, it is rare that a single study has been sufficient to demonstrate the risk of a toxic agent, the benefits of a medical procedure, or the effect of diet or exercise. Some of the studies that have been effective in defining the roles of diet and exercise (etc.) on health have had tens of thousands of participants. Such large numbers are necessary to detect small effects reliably. Thus while one can view the controlled experiment as the ultimate type of test, in practice caution is necessary even here.

Discussion

Anyone who enters a fun house at the carnival is aware that much of what they see will be illusory. As disconcerting as it may be, it is also the case that in the realm of decision making much of what we think is also illusory. If this was not so, we would not find ourselves embroiled in foolish wars, slimy monsters such as Hitler would not obtain power, people would not believe in magic, and those running get-rich-quick schemes would not be so busy. While the human brain is excellent at many tasks, it is not very logical or consistent. It is far too easy to believe what we wish to believe. This is why reality checks are so important. By recognizing where we are likely to be weak, we can compensate with reality checks of various sorts. For various decisions, we must be aware of wishful thinking and self-justification. We must be alert to long chains of reasoning and not trust them overly much. We need to be aware of jumbled collections of unrelated ideas and not give them undue credit by calling them a philosophy. Any time there is risk or chance, we should try to base our decisions on an explicit calculation of probability, as in an actuarial table. We can obtain useful information from reviews by colleagues, while at the same time keeping in mind that a reviewer has all the same limitations of reasoning that we do. Other types of reality checks such as prototypes, tests, surveys, and experiments are also invaluable. Overall, it is very dangerous to proceed without some type of reality check.

2.5   A MATTER OF STYLE

Personality significantly affects problem solving style, the subject of this section. Problem solving style is the overall pattern of how a person tackles a problem, the type of problems they choose, and their emotional relationship to problems. Rather than probing for the deep psychological reasons (motivations) behind different styles, I merely present them here and comment on their utility and limitations for different types of problems. Problem solving style is particularly relevant for complex problems, because certain problem types are only solvable (or even identifiable) using certain styles. Some of the styles we will explore include the fault finder, the visionary, the obsessive, the counterpuncher, the efficiency expert, the synthesizer, and the artist. I conclude with a discussion of cognitive style, as it contrasts with emotional style.

The fault finder

We normally think of fault finding as a negative trait, and when dealing with people it generally is. When dealing with problems, however, fault finding can be a very productive approach. Petroski (1992) points out that most inventors and industrial designers begin with a perception of some fault or deficiency in an existing object, and then devise an improvement to that device, tool, or product. One can not readily improve upon an existing object if one can not see the deficiencies in it. Inventors are typically driven to improve upon existing devices and to create new devices that do jobs that currently no device does well or at all. In this sense they are continuously dissatisfied. Catastrophic failures, such as airplanes that lose their wings and bridges that collapse, have always been a tremendous spur to invention, but even subtle failures are an indication of something that can be done better. The distinguishing characteristic of a successful inventor is that he has a good understanding of what is possible, what the constraints are, and what the tradeoffs are (constraints and tradeoffs are discussed further in Chapter 3). He knows that to make a hand tool smaller but maintain strength may require a stronger, and perhaps more expensive, material. He realizes that to make a tool better for a particular job may make it worse for some other job, which is why combined hammer–screwdriver–paintbrushes have never sold very well. This firm grasp on the tradeoffs, the limits of materials, and the deficiencies of existing objects provides guidance to the inventor so that his changes lead to an improved product, and work as intended. Politicians are similarly always dissatisfied with existing systems, since expressing a desire to improve things generates political support and may even have provided their motivation for entering politics. However, they do not have a firm grasp of the tradeoffs and limitations of their materials, and are constantly trying to control things that can not be controlled and invent perpetual motion machines, which is why the United States (among others) has such a large deficit.

The engineer's drive to improve a product does not necessarily lead to progress. For example, there is a tendency for any new car model to become fancier with each model year as improvements are added, but this tends to increase the cost. It is very difficult for a car maker to resist this tendency. As personal computers have become fancier and more powerful, everyone wants the latest model even if they are only going to send e-mail and write letters. One must wonder, for example, what in the world the typical user plans to do with a 40 gigabyte hard drive unless they are storing movies.

The evil twin of the fault-finding inventor is the nit picker. The nit picker finds fault with everything, but is not able to distinguish the significant fault from the trivial. The typical destructive nit picker is a boss who is never satisfied with a subordinate's work, but typically focuses on the formatting of reports, minor grammatical mistakes, and similar trivia, and may completely ignore the serious flaws in a piece of work. Conversely, he will not praise a good or even an excellent idea or result because there is always something about it that can be picked apart. Such bosses become enamored of word processors and know all the details of formatting, fonts, type styles, and report organization. The consequence completely escapes them that endless picking at and revising a document can make it a year late, and therefore useless.

The visionary

The visionary is a different breed from the tinkering inventor. Once the visionary perceives a problem, he does not seek to improve it incrementally, but to start from scratch to design an entirely new product or system that will be perfect or ideal. The visionary wants to design a car from the tires up, all new and revolutionary. CAD (computer aided design) systems have made it much easier for the visionary to do this. The inventors of the personal computer, John MacReady with his Gossamer Albatross human powered plane, the inventor of the fax machine, Henry Ford with his pioneering assembly line, and Ray Kroc (founder of McDonald's) who pioneered fast food were visionaries. The visionary is driven by images of perfection and by pictures of entire products, processes or systems, not by ideas of tinkering with the tweeters or woofers or making a doorknob easier to grip. A visionary may be very unhappy if he is not in a position to carry out his vision. Many visionaries are frustrated engineers or designers who never get to do a grand project but are instead given pieces of larger projects to do. They can become distressed and disillusioned because their job conflicts with their style. Those visionaries who lack the technical skill to carry out a grand design are what we typically call dreamers: the salesman who designs and sketches cool looking sports cars at home that will never be built, the fringe political types who are sure they know how utopia should be engineered, etc.

Designing from scratch can be an expensive approach in industrial contexts. For example, if every car is designed from the bottom up, then many potentially standard parts such as nuts and bolts not only must be designed, but become nonstandard between models. Further, there is a tremendous overhead to keeping track of 100 different types of fasteners and keeping them in stock. In addition, components that work perfectly well (e.g. windshield wiper motors) when redesigned will typically have bugs at first, thus leading to a defective product and reworking. Thus care must be taken when an employee is a visionary, because his tendency will be to redo everything, including things that don't need redoing. This is even more the case when a visionary type takes over an organization. He wants to shake it all up, reengineer it, restructure it in a completely new way. In a large organization such restructuring is not without costs, both direct and in lost productivity, and may not necessarily lead to improvements. Thus restructuring for its own sake can be seen to be a consequence of the visionary's inner drives and desires, and is not necessarily justified by the bottom line or by any rational plan.

The visionary scientist searches for a grand theory, a unification of his field. Maxwell's equations provided a unification of electromagnetic theory. Quantum theory provided a unification for many aspects of particle physics. Premature unification is also possible, in which an attempt is made to provide a grand theory that is really not successful. Such an inadequate unification can suffer, for example, from vague terms which allow too much wiggle rooms in interpretation, or explicit deductions might not be derivable from the theory so it can't be tested.

The jurist

The jurist focuses on legalistic problems: rules, regulations, permits, procedures, forms, and formalities. This type of person is very concerned with structure and feels very uncomfortable in an unstructured environment. The successful jurist can devise a working procurement system, approval forms, permitting procedures, etc. In this sense they can be successful problem solvers. Jurists are also useful when they can be enlisted to help an organization run smoothly. For instance, a good budget person for a department should make sure everyone gets their financial reports and proposals in on time, head off procurement rules problems, and generally provide preemptive strikes against delays and mistakes. There is an unfortunate tendency, however, for jurists to be so focused on rules and procedures that they lose sight of the need to get anything done. The result, common in government agencies, is approval processes for documents or actions that are arcane and labyrinthine, or that are actually impossible. For example, at one government site one could not send a soil sample out to a lab for analysis to find out what contamination was in it unless you knew what contamination was in it. Similarly, department A may require that department B sign off first, but B requires that A sign off first. A common problem in government procurement is requiring that one spend a month of effort (at more than $12 000/month total cost) to document and justify a sole source procurement of $6000. Bidding requirements may be such that the cost to the bidder of preparing a bid package exceeds the value of the contract, leading to offers with no bidders. All of this results from the narrow legalistic approach of the jurist, who believes that one can pass detailed rules to prevent all possible problems (such as in procurement). This is analogous to stopping and searching every single shopper to prevent shoplifting, or stopping every car and conducting a sobriety test to prevent drunk driving. This view has taken over the United States Environmental Protection Agency and other agencies in their passage of rules to govern very specific behaviors. This view totally misses the cost of regulation and rules. In procurement, the cost of ensuring that every bidder has the ultimate health, safety, and quality assurance plans, complies with all federal labor laws, etc. etc. is that one must pay twice as much (via hidden costs) for products and services. The cost within a beaurcracy of overly arcane procedures and protocols is gridlock. Rules and procedures can help things run smoothly, but when procedural perfection is demanded and procedures are enforced to try to ensure that no mistakes are ever made, then actually getting something done can become like swimming in molasses.

The obsessive

A very successful type of problem solver is the obsessive, not because he has any particular technique advantage, but because he is single-minded in his thinking. The phrase "get a life" applies to most obsessives. A hacker is an obsessive, and thinks about almost nothing but computers. As another example, I know a zoologist who is obsessed with turtles and snakes and another who thinks about nothing but ducks (his house has duck paintings, duck decoys, duck pillows, duck sculptures, etc.). Such individuals somehow have become intensely interested in a particular topic, to the exclusion of almost everything else. This provides them with several advantages. First, they are fascinated by every detail, every dimension of their chosen obsession. Because of this, they know all the trivia and raw facts that on some problems make all the difference, because many times the devil really is in the details. For example, when a computer virus strikes, a hacker is a useful person to have around because they have messed with all the obscure operating system features that are key to tracking down the virus. Second, the obsessive develops an intuitive feel for the object or problem. My zoologist friend thinks like a duck (no he doesn't look like a duck) and has a real feel for their behavior. Another will have a real feel for motors or model airplanes. This intuitive feel is deeper than factual knowledge or book learning about a topic and is very useful for solving problems. Third, because they are obsessed, they think about ducks or computers when they are driving, when they are in the shower, when they are cutting the grass, and thus spend much more time thinking about the topic than other people do.

The obsessive approach to problem solving is not without its pitfalls, however. An obsessive may spend most of his time on unproductive fiddling around, as has been noted for hackers. Obsessives may also be too narrowly focused if the problem involves other domains (interaction of hawks and ducks, computer user ergonomics, etc.). In such cases, the obsessive's overly narrow focus can be overcome by teaming them with someone with other skills, but an obsessive working alone may not realize that they need such collaboration.

The counterpuncher

Muhammed Ali was a classic counterpuncher. When his opponent would swing, he would jab, and gradually he would pound them senseless with his many hits. In problem solving, a counterpuncher is one who reacts to a solution, theory, or idea posed by others. Just as in boxing, the action of the opponent opens up an opportunity. The counterpuncher is typically angered by arguments or ideas that to him seem erroneous or fallacious or is annoyed by products that are ugly or inelegant. The anger that is provoked is an indication that the person somehow recognizes that the argument of the opponent is in error, though they may be unable at first to say what the error is. This anger simultaneously provides a motivation to prove the other person to be wrong. The philosopher Kant was angered by the (to him) fallacious arguments of Hume, and set out to build an entire new system of rational philosophy to counter Hume's "false" doctrines. The counterpunch style can be seen in the development of the Apple computer, which was a reaction to the mainframe way of thinking.

The efficacy of the counterpunching style depends on several factors. First, the person must have a good instinct for what is worth getting angry about. If they get angry about silly ideas like astral projection or healing crystals, then they will not be productive. Second, they must have the technical skill to convert the anger into a productive alternative. Finally, they must not become so angry that their judgment becomes clouded. Karl Marx, for example, was so angered by an unjust system (nineteenth-century Capitalism), as well as being angered by being an academic outsider, that he went to extreme lengths to prove his thesis, thereby losing all regard for facts and logic (Johnson, 1988).

The efficiency expert

Some people have a fascination with problems of efficiency. They love the challenge of finding a faster, easier, or cheaper way to do something. Since real world problems often involve efficiency considerations, the contributions of the efficiency-minded problem solver can be valuable. We may wish to reduce electrical consumption in a refrigerator or speed mail delivery, reduce waste in a factory, or shorten the checkout lines at a grocery store. The efficiency-minded person is inherently irritated by material waste, inefficiency, and wasted time. Thus they notice waste more than others and are motivated to find ways to reduce it. Efficiency improvements are most valuable in the context of existing products or services. Federal Express is an example of a company built around a concept of efficiency, but it required a visionary to create the company, not an efficiency expert alone.

Efficiency experts can cause difficulties in certain contexts. New product development, for example, is risky and is therefore wasteful. Such messy, inefficient operations drive efficiency experts up the wall, and are likely to be stifled by them if they are in charge because they will tend to request proof that any given action or expenditure will produce results, such proof of course not being available. They also become impatient with the slow and tedious R&D process and may cut projects off prematurely. Finally, an efficiency focus can lead to short-sighted behaviors. During the rapid growth phase of a successful new company one should not try to optimize every process because they are likely to change rapidly. Further, bean counting can take effort away from high-profit activities and growth in favor of marginal returns resulting from cost-savings efforts or efficiency considerations.

The tendency to push a system toward more and more efficiency may make it brittle. For example, the perfectly efficient assembly line does not allow for workers to use the bathroom or take into account equipment failures. The perfectly efficient business does not have any slack for rush jobs, and will therefore lose some business and maybe lose customer loyalty. This is the fallacy of the downsizing fad: on an economic downswing it always appears that there are excess personnel but as soon as the economy picks up a little, those excess personnel become essential. Thus companies end up in a cycle of hiring and firing that is destructive of morale and not even optimal for profitability because hiring is expensive and new hires are not that productive at first. A very thermally efficient shower stall is available that has recoiled tubing in the floor that acts as heat exchangers to capture waste heat in the drain water, but it costs $2500 and if the piping in the floor becomes clogged with hair it can never be roto-rooted out. Some amount of chatting around the water cooler is essential even though inefficient because it promotes alliances, networking, and exchange of information. Thus the tendency to view efficiency as virtuous and essential must be resisted, because efficiency is only one of the goals to be achieved in any problem context and because efficiency and growth are not necessarily compatible.

The synthesizer

A very particular skill is possessed by the synthesizer who relishes drawing all the threads of a problem together and weaving a comprehensive, integrated solution. He pulls it all together to write the definitive textbook on a subject. The best biographers and historians are synthesizers (the worst being mere catalogers and list makers). If one wishes to trace the consequences of a foreign policy action, a synthesizer is needed. The synthesizer wants to arrange all the pieces of the puzzle on the table and manipulate them until they all fit together. If they can not be made to fit, then the synthesizer may be working on a problem prematurely, and is at risk of achieving a false synthesis. Karl Marx was a synthesizer as well as a visionary, but pursued his vision by forcing facts into place with distortion and fabrication (Johnson, 1988). When the facts do not all fit together, however, what may be needed is a tinkerer or visionary who can redefine the problem. Synthesis requires time to contemplate, to gather facts and sift them, to formulate overall structures and explanations, and is thus difficult to do in our fast paced modern world. Even when a synthesis is achieved, it may be difficult to communicate to those who only have time for one paragraph summaries because a true synthesis does not fit into a bulletized format.

The tool master

The tool master has mastered one or a few tricks or techniques. Being a man with a hammer, everything looks like a nail. He is the person who can do anything with a spreadsheet, but not much else. This is the person who wants to computerize everything, all data, all communications, everything, because that is his tool. Tool masters become most visible when there is a clash of technologies: those of this type wedded to old tools and techniques will be in tooth and nail conflict with those of the same type (usually younger) who have adopted the new tools. The difference between the tool master and the obsessive is that the obsessive is in love with a subject (turtles, cars, the Civil War) whereas the tool master has hitched his professional wagon to being a competent user of some tool or technique. Such a person can command high wages if the tool they have mastered produces value and is in demand. One can observe many such people in academia who have mastered some technique (toxicity testing, gene splicing, literary deconstruction) and apply this technique to one problem after another. That is, such people are not subject matter experts but are experts in technique. The problem with this strategy is that real problems may have many dimensions and may require many tools for their solution. Thus the tool master may be restricted in what he can accomplish to those problems that fall within the scope of his bag of tricks.

The artist

Those whose problem solving style is governed by aesthetic sensibilities are not all artists in the traditional sense. Rather, they are driven in their search for problems and solutions by aesthetic criteria such as symmetry, harmony, elegance, beauty, and simplicity. A mathematician, for example, may be most attracted to an elegant proof, and an inventor annoyed by a clumsy piece of machinery. A good aesthetic sense is very difficult to describe, and even more difficult to inculcate in a student. In ancient Egypt, artisans were trained from a rigid manual that prescribed exactly how human figures should be drawn. Little aesthetic sense was developed from such training and all artists' work was nearly identical. In modern art, it almost seems that anti-aesthetics governs because the motivation is often to shock, to be different, to irritate, or to make some social or political statement.

Nevertheless, when designers and inventors are guided by a search for beautiful or elegant solutions, they are often on the right track. This is because beauty is often consistent with such useful design principles as efficiency, balance, and utility. For example, if an electronics board has wires running all over the place, creating an ugly appearance, then it is probably badly designed and will be both slower and generate more heat than a better design would. An elegant mathematical proof is often more powerful and general. An awkward computer operating system will be hard to learn and the user will make many mistakes.

The recognition of beautiful patterns is often the first step toward the discovery of simple or interesting organizing principles. The shapes that soap bubbles make on wire frames are directly due to least energy principles, as are the shapes that crystals take. The haunting beauty of the Mandelbrot set galvanized interest in fractals. Symmetry of animal body form has been found to be upset by deleterious mutations. Our perception of lush greenery somehow relates fairly well to the absence of soil mineral deficiencies. Thus an intuitive sense of harmony, symmetry, and elegance can be a very useful guide to discovering or creating.

Aesthetic considerations are not a foolproof guide to one's work, however, because they can be influenced by fads. Automobiles have been influenced by various views of what is aerodynamic, with a common "low-drag" car (e.g. many sports cars) being tapered in front and squared off in back. However, since the most drag is created not by the nose but by the creation of turbulence at the rear, such cars would actually have less drag if driven backwards. Our aesthetic concepts of what is modern or space age often conflict with what is efficient or useful, which can be seen most clearly when looking at old magazines purporting to show what the future would look like (way off in the 1980s or 1990s!). Pursuit of an aesthetic theory can also mislead. Many ecologists have utilized overly simplified mathematics for populations dynamics because it leads to elegant (aesthetic) solutions. More realistic (and not so pretty) representation of dynamics leads to different conclusions.

The craftsman/craftswoman

The craftsperson, a dying breed, is concerned with the total product and wants every aspect to be perfect. The product is not necessarily an invention, and may be something that is made the same by others. Traditional craftsmen/women made things by hand and took pride in their work. As used here, the craftsman/woman is a person concerned with the aesthetics of the product, its functionality, and with the details. If a person with this style is a scientist, they do the experiments themselves rather than delegating the work to students or a technician. If the person with this style is a store owner, the store is not just a source of income, but is a source of pride. This store owner will be found tidying up and making the store look good, helping customers, and answering the phone himself. Certain consumer products have the feeling of having been overseen or designed by a craftsman/woman, because there was great attention to overall functioning and the integration of every detail.

Other products, in contrast, look like a different person designed every part and then they were all bolted together.

The craftsman/woman will have trouble doing a rush job or creating a rough prototype of a product. They will tend to tinker too long, polish too much, and worry about style and format on projects that do not warrant such care. It is important to know when detail and polish are warranted and when they are not.

The eccentric (the genius)

It is commonly believed that really groundbreaking work requires that one be an extreme personality, like Van Gogh, and cut off one's ear, or drink too much like Ernest Hemingway, or have long hair and wear wild clothing. That is, that one must be an eccentric, that genius is inextricably linked with manic depression, with melancholy, with flamboyance, and with deviance. The facts, however, do not support this view (Arieti, 1976). This is a romantic notion that results from an extreme bias in historical and popular reporting such that those individuals who are both innovative and deviant receive undue attention because they are the most interesting. Einstein was not the only prominent physicist of his day, but he created great photo opportunities, spouted popular philosophy, wrote letters to presidents, and had endearing eccentric traits such as getting lost on his way home from work and never combing his hair. Richard Feynman is another such wild and crazy guy. Such people make good copy, and in particular fit the stereotype of many writers that the genius must be eccentric. This is extremely misleading. First, some of the most prolific people, and the greatest minds, have not been written about at all. Frederick Sanger, one of the few people ever to win two Nobel Prizes, is absolutely normal, modest, and humble (Jones and Douglas, 1994), and consequently is virtually unknown. There are inventors who have changed our technological world almost single-handedly (Petroski, 1992) but whom no one has heard about because, frankly, they are boring. The high percentage of literary prize winners who are depressed, disturbed, or suicidal similarly results from the bias in literary prizes toward those who write disturbing, moody, or depressing works. Those who write best sellers, detective novels, or science fiction novels rarely win such prizes. Second, I believe there are far more eccentric people who never achieve anything than there are high achievers who are eccentric. There are thousands of UFO enthusiasts, flat earth aficionados, collectors of 10 000 hubcaps, and Civil War buffs who squander their enthusiasms on the trivial or the fantastic. The eccentricity of these individuals does not help them solve real world problems. Further, a focus on personality, particularly a focus on the roots of extreme behaviors, the neuroses of the great creators and their obsessions, does not help us uncover the problem solving techniques they used (Arieti, 1976), techniques which may in fact be useful to more well-adjusted individuals as well as to the eccentric.

What we can say is that in some circumstances the eccentricity of an individual may help them if it allows them to come to a unique perspective or to be oblivious to the fact that they are out of step with the standard way of thinking. In particular, those in the creative arts tend to be much more likely to be afflicted with mood disorders, particularly manic depression (bipolar disorder), which may help them to be more expressive than they would otherwise be since they tend to be flooded with unbidden images, word associations, and moods, particularly during their manic phase when their level of creative output may be enhanced (Jamison, 1995). What seems most likely to me is that when high intelligence is combined with a mood disorder, the individual in question feels compelled to turn their energies toward artistic creative expression and that conversely such a person would not do well in science courses. Since such expression need not be functional, they may be successful in such efforts. We may note, however, that those with mood disorders, and particularly manic depressives during their manic phase, exhibit extremely poor judgment with respect to real world affairs, tending toward extravagant schemes, unrealistic plans, and impulsive behaviors (Jamison, 1995). This is why no such correlation of mood disorders with success in science or in business can be found: in these domains steadiness and clear thinking are essential.

It is quite possible to be normal and to be brilliant. In particular, this book is based on the proposition that whatever level of ability one has can be used more effectively. In fact, it is likely that almost anyone can be twice as effective to an order of magnitude more effective (see Gilbert, 1978, for documentation) if they can learn to (1) use their mind and creative abilities, (2) not work against their style (discussed in this section), (3) learn to use their time effectively, and (4) learn the techniques of strategic problem solving. Thus we should resist the temptation to undervalue our ability because we don't appear to be an extreme personality such as Freud or Picasso.

Discussion

The suite of problem solving styles discussed above is not exhaustive. The noted problem solving styles are also not personality types, and an individual may use different styles at different times or in different domains. Some flexibility in this respect is a good safeguard against the potential flaws inherent in each style. This examination of style explains a long-standing puzzle in studies of creativity. There exists an extreme disparity between the personality traits assumed by different psychologists to be conducive to creativity (Arieti, 1976; Sternberg, 1988). Some assert that the innovative person is flamboyant but others that he is withdrawn or introverted, some say aggressive and others say detached. By examining style, we can see that much of this disparity results from a failure to characterize the type of problem the persons studied were successful in attacking. It is not the case that creative persons are creative in all domains, nor that their approach to problem solving will be universally applicable. What we can say is that those noted for their accomplishments have been successful in solving some type of problem or creating some type of product, but if we characterize their work it may often turn out to be dominated by a particular type of problem or to have been created in a certain way. That is, the effect of personality is to influence problem solving style, which affects problem choice, method of attack, emotive content, problem scope (tinkering versus reengineering), and other factors, and thus the domain within which the person will be successful (or not). There is no single creative personality.

A further consideration is that problem solving style may interact with intellectual talent or special strengths. Howard Gardner (1983), for example, believes that there are seven distinct types of intelligence, including linguistic, mathematical, musical, spatial, interpersonal, and others. In this book it is argued that strategic intelligence is a further basic skill or ability. An IQ test may be a highly inadequate tool for assessing mental abilities. Individuals are likely to have ability in these different domains to differing degrees. Degree of talent in an area, of course, is relative to how much training one has obtained, so that math phobic individuals may actually have considerable native mathematical ability, but had negative experiences when young. Nevertheless, to the extent that these particular dimensions of intelligence (native and/ or polished by training) coincide with the style of the individual, then that individual will be more successful (or not if not). For example, someone with a strong spatial ability and a visionary style could be successful as an artistic designer, but not necessarily as an industrial designer. Some individuals have strong aesthetic motivation, but do not necessarily have much talent in this area (e.g. no ability to visualize, no spatial intelligence). Such people may become collectors of music or art or become critics. Some people are fortunate enough to be able to overcome mismatches between their problem solving style and their intellectual strengths by forming partnerships with others. Such teaming is common in the arts, where singer/songwriter pairs often become inseparable. Teaming is also common in the sciences.

Overall, problem solving style has a significant effect on the work individuals choose to do, how they go about it, and how successful they are. Differing styles between managers and subordinates or between project team members are typical causes of workplace conflict. Style is a very significant factor in problem solving, a factor that is often overlooked.

2.6   ATTITUDE: THE INNER STRATEGIST

The purpose of this section is to explore the role of attitude in enhancing strategic problem solving and innovation. Attitude is well known to be important in sports, sales, and the arts, but is not as appreciated in the context of problem solving. However, because our thinking processes are even more subject to derailment, delusion, and mistake than are our aesthetic and athletic performances, the role of attitude, the inner strategist, is even more important. As a model for assessing the role of attitude on problem solving, I take not the arts, but science. The arts are primarily a vehicle for emotional and intellectual expression rather than practical problem solving. Furthermore, in the arts no objective criteria exist for judging success or truth. Thus if an artist attributes his creativity to drugs, it is very difficult to evaluate this claim. In science, the creative act must result in a practical, workable solution to some problem. There is an objective criterion for judging success in science by which we can evaluate whether some attitude or behavior helps or hinders creativity. It would also be possible to use engineering as a model here, but there are too few accounts on which to base an analysis.

The creative act

Not everything done in science may be considered creative. A major part of the edifice of science consists of a toolbox of techniques and instrumentation. We are interested here in the subset of scientific problem solving that we may label truly creative. Such work is easily recognized in retrospect; virtually every scientist who makes it into the history books has done creative work (though not all their work would

qualify). The historical perspective selects those who were both creative and correct, though we must note the historical bias toward colorful individuals that taints this source of data. It is much more difficult
to evaluate creative work as it is being produced. There are many
characteristics shared between the truly creative and the misguided.
Bold hypotheses, eloquent presentation, and beautiful analogies characterize both groups. What separates the crank (such as those who
continually invent perpetual motion machines) from the true innovator, in my opinion, is a fatal flaw in a key behavioral dimension of the
creative process in the crank. Such flaws include a lack of humility, self-
aggrandizement, lack of introspection, and dishonesty. Please note
that I am not suggesting that great scientists are saints. They tend to
avoid church-going and have little interest in participating in charita-
ble organizations (Simonton, 1988). However, the focus caused by work
that is reality oriented and provides constant feedback leads to certain
positive attitudinal consequences. It is impossible, for example, to
conduct research while hung over or in a rage. Most creative scientists
are particularly conscious of the factors that interfere with their work.
Likewise, because the creative act is subtle and easily disrupted, the
connection between inner realities and concrete results is perhaps
more easily demonstrated in the context of science where objective
testing of results is possible.

Hubris and humility

Making creative contributions in science requires a delicate balance
between hubris and humility. Hubris is defined as insolence or arrogance resulting from excessive pride. It is generally a pejorative term,
but in fact insolence and arrogance (of a special kind) are essential
qualities for making truly innovative discoveries. The creative innovator is insolent, having little respect for authority, because his experience is that authority is often wrong. In fact, his career depends upon
being able to show that the current wisdom is wrong or at least that it
has overlooked something important. He is arrogant because he
expects to be able to uncover the secrets of nature by his own power,
and in fact makes a career of doing so. This is hubris in the sense
exhibited by Prometheus who stole fire from the Gods, though it is
often viewed by outsiders in purely the pejorative sense. The original
thinker must endure periods of solitude and must expect to be mis-
understood and perhaps bitterly opposed by peers. The ego must be
strong to withstand these hardships. Thus, while ordinarily we would

consider hubris to be a negative trait, in the case of the explorer or innovator in any field the above argument indicates that it may be an essential quality for success.

The key to the successful use of "hubris" is that it be tempered by humility. Hubris tempered by humility leads to a balanced type of confidence that we may label strength of character, for want of a better term. The difficulty here is that even if someone is truly humble, the mere fact that they are doing or attempting something great or difficult leads others to view them as arrogant or prideful out of jealousy. Jealousy among academics is rife and can be bitter. Thus those who dare to achieve greatness and challenge authority by discovering new things risk the label of arrogance and pridefulness due to the jealousy of others, even when the charge is unjustified.

Several considerations can help one achieve humility. First, one can, of course, make mistakes and thus must be careful not to state too strongly that one is right and others are wrong. Further, the fact that one's work supersedes previous work does not make one's predecessors stupid. Second, most discoveries, systems, inventions, and theories are eventually superseded by better, more precise, more comprehensive ones. In rapidly moving fields (e.g. computers) this can happen within a few years or less. One must dare to have a grand hypothesis but remember that it may well be soon forgotten. If one has as a primary goal of one's work service to all humanity, then the product of one's labors becomes a gift to be given rather than a status symbol or a bid for immortality. When an individual desires success not merely for the sake of achievement but to show that they are really better than or superior to others, then this is the death of humility. That is, one must ask whether the climb to the mountain peak is for the challenge and the joy of achievement or to make one feel taller.

Failure to balance these factors properly is at the root of both underachievement and barriers to strategic thinking. Lack of confidence leads to underachievement because the scientist refuses to speculate. Without bold hypotheses, models, and experiments one is reduced to plodding routine. Such attitudes lead to a desire for absolute certainty and a tendency to dismiss anything new as "mere speculation." Failure to understand that the intellect is far more powerful than we assume also leads to confusion between one's self-esteem and one's actual abilities. Those whose self-image is damaged by events in their lives or who feel themselves to be young and inexperienced will fail to take seriously their true level of talent. They will then not attempt the "difficult" or important problems. But many problems in retrospect are seen to have simple solutions or to have yielded in the end to pure persistence. Many of today's key problems are as open to solution as more mundane problems with far less payoff. Thus the lack of hubris causes a great deal of underachievement.

Hubris unbalanced by humility leads to several types of sins of excess. The most visible problem is what might be called the crackpot syndrome. When a brilliant and successful strategist becomes convinced of his or her own freedom from error or the certainty of his or her particular hypothesis, he or she can become immune to criticism or to contrary evidence. This leads to many endless rounds of debate in academic fields because no amount of evidence is sufficient to change the mind of the fanatic (Loehle, 1987). In political fields it leads to great disasters (the invasions of Russia by Napoleon and Hitler). In business it leads to overly rapid expansion of a new enterprise or the entry by a business into a new venture about which it knows nothing. A tricky aspect of evaluating particular cases is that many who seem to be crackpots at the time turn out later to be right. When Wegener proposed the idea of continental drift 90 years ago this must have seemed exceedingly crazy to his contemporaries (he was in fact vilified for his views) and yet he was essentially correct. There is a key difference between his behavior and that of a crackpot, however; Wegener was not immune to criticism and fully acknowledged that a weakness of his theory was the lack of a mechanism by which continents might move. Thus we can see that he was ahead of his time, but not suffering from irrational adherence to his hypotheses. Darwin likewise was tenacious in his views but was also acutely aware of the mechanisms and facts that were lacking to establish his theory of evolution. Darwin in fact spent many years attempting to fill in the gaps in his theory. The extrapolation of these examples to other realms of endeavor is obvious, however the farther one moves away from the hard sciences the harder it may be to tell who is really a crackpot and who is a visionary.

Far more common than the crackpot is the "expert." The expert has also forgotten all about humility and is very proud of the fact that he knows all about his chosen specialty. The expert criticizes new ideas because they contradict what we "know" to be true. The expert is frequently also intolerant of those who intrude on his intellectual territory. Even worse, becoming an expert actively interferes with creativity and discovery, as discussed in Section 2.1 "Strategic creativity." Failures perpetrated by "experts" include the complete inability of mainframe computer makers to see the future strength of the personal computer market and the frequent rejection of new products and inventions within a corporation by the senior brass. Humility constantly reminds us that the certainty of yesterday will be completely toppled tomorrow; thus we should be ready at any moment to discard any piece of knowledge and start afresh.

Detachment

A key to maintaining the proper balance between "hubris" (or confidence) and humility and avoiding the pitfalls inherent in any type of creative or innovative work is a spirit of detachment. Detachment is a rather alien concept in current Western culture which vaunts the self and personality above all else and which holds self-esteem as the highest good.

Detachment is an attitude that arises from the realization of several key truths. The first of these is that our true worth is unrelated to our material attainments. When one realizes that the most valuable things in life are the quality of one's life, the strength of one's character, and one's family, then one is on the way to true detachment, because these most valuable things do not accrue from material success nor can they be taken away. They are entirely internal and are not subject to success or failure or the opinions of others. In any problem solving domain one's accomplishments and discoveries are quickly obliterated by new discoveries and better products, so the ephemeral nature of material accomplishments is more obvious than in certain other fields of endeavor.

A spirit of detachment is a great gift in the pursuit of the creative solution of problems. One of the most important characteristics of creative individuals is independence of thought (Root-Bernstein, 1989). To be truly creative or innovative one needs first to have a mastery of the tools of the trade, but then one must go beyond what is known and create something new. Detachment enables one to recognize that the current state of knowledge is ephemeral and not fixed in stone. Furthermore, by removing excessive concern over the opinions of others, one's mind becomes free to consider new ideas clearly and on their own merits. New ideas are fragile and easily destroyed by premature criticism. Detachment leads to the kind of objectivity that can nurture truly original ideas.

Detachment is also a mighty shield against the slings and arrows that face the would-be innovator. Detachment shields one from fear of failure and fear of criticism, the two big fears that cause many to shy away from the new or unusual. True innovators are bound to seem odd or to receive unwarranted criticism, and detachment can help deflect negative responses without the individual's needing to isolate himself or herself or develop a bitter, defensive persona in response.

Conversely, detachment helps combat the rigidity, conservatism, and know-it-all-ism that tempt the expert but kill innovation. Since the knowledge of even the most brilliant person is but a drop from the ocean of all knowledge, and since all around us the unknown far outweighs the known, the idea that one could ever truly "know it all" or be an expert can be seen clearly as an exercise in self-delusion. The extreme forms of hubris that result from the identification of one's self as the owner of an idea (and thus the crackpot's refusal to change his mind) are likewise tempered by detachment. Not only does one see then that one's discoveries are not really due to one's own efforts (because one's very skill is a gift), but one sees that these products of creativity do not belong to oneself at all but are truly gifts of service to humanity.

Tranquility

In our increasingly frenetic world, tranquility is a scarce commodity. It is not merely our nerves that suffer, however. Lack of place and time for reflection have an adverse impact on our inner lives. The need for outer tranquility in the search for inner tranquility has a long tradition and is institutionalized in various forms of retreats. What is not so well known is that inner tranquility is also conducive to success in creative endeavors, including the type of strategic thinking discussed here. Consider the sport of orienteering in which one attempts to navigate across country using a map and compass. The winner of an orienteering race is rarely the swiftest runner, but the one who has his or her bearings. In complex problem solving also it is crucial to have one's bearings. The ability to still the mind, to achieve a calm, reflective attitude can contribute significantly to innovation and creativity. This is because the most important and yet most difficult step in solving a complex problem is the formulation of truly original ideas. Yet good ideas typically start as urgings, hints, wisps, vapors, images, or vague analogies. Only the rare individual has them pop up in a concrete form, ready to act on. More likely is that the phantasm will need encouragement and the patience to watch it drift by at its own pace. A hurried, impatient, "busy" person will not give these ideas sufficient play or attention and will tend to rush on to more "productive" work. Whereas the painter typically works alone and is almost forced into some sort of meditative or introspective state, a state in which verbal modes of thought are suppressed, it is all too easy for the scientist to rush from meeting to airport to computer, and the phone is an all-too-frequent interruption. A frenetic pace is very detrimental to the formulation of complex thoughts and receptivity to nascent breakthroughs.

Tranquility, however, is more than just a technique such as walking or the safety of a remote office, free from phone calls. One can take a quiet walk in a remote wood yet still find one's mind racing with thoughts and anxieties. True tranquility is brought about by the cultivation of certain key attitudes. First among these is detachment. Constant dwelling on getting the Nobel Prize or getting that next promotion poisons the intuition and receptivity to new and subtle ideas. The realization of the smallness of even the greatest of our accomplishments allows one to maintain a degree of emotional distance from the work at hand so that the needs of the work itself provide the guidance for how to proceed rather than our hopes and wishes. Detachment also allows ideas to enter our consciousness that may contradict our previous results (and perhaps contradict our cherished hopes), which a busy and emotional mind will suppress or ignore but which are crucial to correcting one's course. Tranquillity may also be enhanced by practice of various disciplines such as meditation.

Honesty

One might not think of honesty as relevant to strategic problem solving, but in my experience prominent scientists are quite honest, both with themselves and with others. Note that this is not necessarily true universally, because some reach prominence due to the labors of subordinates or by being prolific at the expense of being innovative. Why might this quality of honesty be important? After all, problem solving has never been considered a domain for ethical difficulties, as are law or politics. The reason is that if one is to successfully confront complexity and create innovative products, designs, or solutions, absolute lack of self-deception is required. One can deceive the customer, bluff in sports, and bully in business, but nature can not be tricked. When one builds a space shuttle or designs a computer system, the outcome is not influenced at all by one's bluffing or charm or good looks. On the contrary, self-deception is devastating in such a context. There is a natural psychological tendency to seek evidence that confirms what we already believe and to ignore evidence that contradicts our beliefs (Loehle, 1987). Collecting data that can only confirm what we already believe is useful as a check on knowledge acquired by others but it does not lead to progress. Thus one must constantly ask oneself "Am I interpreting this data according to my preferences or am I really seeing what is there?" Failure to be honest can lead to tragic consequences. For example, a prominent turn-of-the-century geologist became convinced in his later years that all rocks were fossil deposits of aquatic microorganisms. His self-deception was so complete that no matter what type of rock he looked at under the poor quality microscopes of that day, he saw microfossils. During the last years of his life no one would publish his work because it was so obviously biased and immune from criticism. Another example concerns the O-ring failure that caused the space shuttle to explode. Even though engineers reported this problem, management glossed over it and refused to deal with it. Self-deception and wishful thinking are rife in business and lead to many failures (see also LeVay, 2008).

It is well to remember that slightly crazy ideas are crucial for successful strategic problem solving. The wild analogy or model is often useful or even correct. Those who become passionate advocates of such ideas are thus not necessarily deceiving themselves. It is instructive to examine how such people deal with facts contrary to their beliefs. When the first experiments testing Einstein's theories seemed to contradict them, Einstein insisted the experiments were wrong. He had reason, however, to be skeptical because of the power of his theory and the difficulty of the experiments. Consider the case of Pasteur (Root-Bernstein, 1989). He had a set of grand hypotheses or goals that in retrospect we can see to be not only wrong but a little crazy. For example, he dreamed of creating left-handed organisms (constituted from chirally left-handed organic chemicals) and thus become famous for creating a new form of life. In attempting to prove his theories he came upon many experimental results that were not what he expected or what he was looking for. He had the scientific honesty (or perhaps objectivity) to confront these results for what they were and publish them. Other scientists of his day had previously observed many of the same phenomena but failed to believe them because they contradicted what they already "knew" to be the case. Thus in the case of Pasteur his honesty saved him from becoming a crackpot (though of course, he was not a saint nor even always honest). It is thus clearly important to be honest with the facts and not let one's biases lead to self-deception. No one is ever free of biases, but honesty about their existence can prevent their negative effects.

Conclusion

I hope to have demonstrated that how one works is not unrelated to one's attitudes and behaviors. Not only do attitudes enhance problem solving, creativity, and productivity, but it is doubtful if one's full capacity can be achieved when negative attitudes intervene. Furthermore, serious dangers await those brilliant minds who attempt to scale great intellectual heights without the protection of humility, detachment, and honesty. By harnessing positive emotional states we risk less and can achieve more than we would otherwise dare to imagine.

2.7   STRATEGIC THINKING EXERCISES

One might ask what exercises can increase an individual's strategic thinking ability, besides reading this book, of course. There are in fact a number of concrete skills, practices, and techniques that can substantially enhance one's abilities in this area. It was mentioned early in the book that intense concentration is essential for success on strategic problems. Games such as chess and problems in math both help foster this ability to concentrate. Thus, experience with chess and math carry over into other domains, even though the specific skills involved may not.

A key difference between the thinking of an untutored individual and a practiced strategist is that the strategist has mental structures at hand for organizing information and connecting events, whereas the untutored individual has disordered thought processes. The simplest example is the ability to order things in time. An impulsive person takes actions without following out the sequence of what will happen next, and is thus always being surprised by consequences. Such a person does not schedule their errands so that their driving time is minimized and so that they finish in time, but is likely to go to the store and then find that their ice cream is melting in the car while they stop to shop for clothes. An impulsive person starts a major outdoor yard project 30 minutes before it gets dark, and ends up trying to cut the grass in the twilight, thereby running over the hose with the mower. A simple strategic structure in this case is merely the ordering of a sequence of events such that the time each takes is accounted for and the tasks fit in the time available. This is a linear structure. The many courses offered on time management suggest that sequential scheduling is not an inherent skill for most people, but must be learned.

As we will see in the chapters that follow, skill in scheduling is an essential requisite to successful strategic problem solving. For example, the generation of scenarios is dependent on sequential scheduling ability. This ability is most easily learned by example. Graduate students who work for a professor who is skilled in this area can pick up the feel for ordering tasks in time so that time is not wasted. Working on a construction project also provides experience in this area as well as providing many examples of problems caused by scheduling failures (supplies that have not arrived, crews getting in each other's way, etc.).

A second type of structure is hierarchical. The organization of a textbook is hierarchical, because there are headings within sections within chapters. The practice of writing term papers in school helps teach this type of organizational skill. Another very useful discipline for learning hierarchical structures is computer programming. By this I do not mean the use of a spreadsheet or a word processor, but procedural languages such as FORTRAN, Pascal, or C. In these languages, certain tasks are compartmentalized in subroutines, and subroutines arranged into a structure according to the computations required. Training in computer programming can help one learn to organize complex tasks into a structure. Programming also has the advantage that one can incrementally improve the program until it works, thus learning how to track down problems and improve a structure.

More complex structures are not necessarily hierarchical, but are often represented by various types of flow charts. Learning to read maps, blueprints, and diagrams is good training for these other types of charts. Those who routinely use such charts may be surprised to discover that many people have great difficulty with maps and diagrams. Many people, for example, have trouble reading a road map. This trait is somewhat gender specific (Gardner, 1983). When giving directions or finding their way, for example, men tend to use ordinal directions and distances as a guide, as if they were looking at a map. Women, in contrast, tend to focus on landmarks. Similarly, there are differences between individuals in their ability to visualize a map or chart. Those with a visual-type memory have an advantage here, although they must still learn to relate the map abstractions to physical landmarks and directions. Those without the ability to memorize visual information may be able to learn to read a map but can not necessarily pull it up for reference mentally while driving.

Charts that capture more complex structures include topographic maps, project flow charts, organizational charts, decision trees, simulation model structure diagrams, and software flow charts. All of these types of charts capture a great deal of complexity in a single diagram and are thus extremely useful. Becoming familiar with these types of charts also provides tools for the internal organization of information and for planning.

Military campaigns provide invaluable examples of strategic reasoning. There is a goal to accomplish, risk and uncertainty, tactics for reaching the goal, constraints, and obstacles. It is very useful to read about such campaigns because we can follow through the consequences of the strategic decisions that were made, particularly in campaigns that are well documented such as much of the Allied effort in World War II. The small-scale model for war, of course, is team sports. It is often held that team sports provide good training in strategy. One must size up one's opponent and assess what they are likely to do, assign tasks to members of the team, who in turn must keep in mind not only the overall strategy and their part in it, but also the likely actions of their own team members so that they can interact properly. Board games also provide training in this type of strategy, though in a more restricted domain. Chess is held up as the ultimate strategy game. Experts in chess learn to recognize strategic patterns in play.

It is true that team sports are good training for some aspects of military type strategy. In real life, however, the "game" is often much more similar to Calvin ball than to a traditional game. This is the game played by Calvin and his toy tiger in the comic strip Calvin and Hobbs in which the rules change constantly. In business settings the rules also can constantly change as one's opponents invent new technologies that make one's products obsolete. A new advertising strategy can cleverly cause consumers to focus on elements of the product that were never considered relevant before. This element of rules that change constantly also occurs in warfare (the winners usually have changed the rules, as when Hitler's troops introduced the blitzkrieg tactic), but does not occur in sports or games, which is one reason they are really rather unrealistic. That is, a good game player does not necessarily make a good life strategist.

It is well to remember that not all strategic problems involve an opponent, as do board games, sports, and military campaigns. Inventing a device or product, planning a trip, moving to a new house, and investing are all strategic problems but do not involve an opponent (unless of course you believe that life is your opponent and is out to get you). In personal strategic action, one's opponents are fuzzy thinking, inertia, self-deception, disorganization, lack of information, and lack of requisite skills. One must obtain and organize information, come up with good ideas, organize components of a task, and be properly motivated. None of these components of strategic problem solving are enhanced by board games, or particularly by video games.

In many strategic problems, there is an element of risk or uncertainty. A business we invest in may fail, the stock market may go down, the weather may spoil our camping trip. As mentioned earlier, people are extremely bad at figuring risks or chance events. Unfortunately, the only solution to this problem is specifically to become proficient at calculating odds. One type of odds is given by games of chance such as cards where there is a known domain. In this arena, one can figure out exactly how likely any given hand of cards is. The same holds for coin tosses or dice. Familiarity with such problems is useful in other contexts also. For example, the odds on having a boy or girl, or a given number of each out of x children is exactly analogous to a coin toss, and can be figured in the same way. For other types of risk, it is useful to become familiar with historical odds, such as for the actual risk of death from automobile accidents or plane crashes. For such data, it is useful to look beyond the initial figures and examine a breakdown of the numbers. For example, one's risk from automobiles may seem high, but a large percentage of total automobile deaths are very late at night and result from fatigue, or are solitary drivers who have been drinking and hit a stationary object. If one neither drinks nor drives late at night, then one's risk is much reduced. To take another example, the risk of small business failure is often given as 50% in the first two years or something similar. What is not stated is that "failure" is defined as the number of businesses that are no longer in business after a given period. However, many of these businesses may no longer exist because they were sold to a larger company (death by assimilation), because the owner died or retired (the eventual fate of many sole proprietorships), or the owner made a change in career. In none of these cases is it necessarily true that the owners lost money or "failed." To properly deal with these kinds of odds-based issues, it is crucial that one have experience with calculations of risk and probability.

Logistics is a useful domain for learning about organizing complexity. In a typical logistics problem, one must get the right materials to the right place at the right time, under various constraints and risks. Not surprisingly, battles have been lost because of bad logistical operations (at the Battle of the Bulge in World War II, the Germans ran out of fuel). Conducting a logistical operation provides good training in various aspects of strategic thinking. Of course, few of us will be in charge of a military supply operation, but there are other situations that can provide similar training. A simple and accessible one is preparing for a backpacking trip. One must bring not merely food containers (cans, plastic pouches) but a means to open them and a way to carry the waste home. One must balance the risk of rain against the extra weight that rain gear entails. One wants to bring just enough food (because it is heavy) but never go hungry. Items must be packed so that snacks and sweaters are accessible while hiking without unpacking the whole thing. This type of problem is analogous to other strategic problems where the limiting factor may not be the weight one can carry but the amount of time available or the amount of money one has to invest.

A surprising type of useful exercise comes about from being a critic. When one volunteers to be a critic of someone's project or is on an official review body, this provides an opportunity to check another person's work for flaws and goofs. It can in some instances provide real insights into how someone else thinks if you can find the flaw in their reasoning. This experience is useful because once a habit of being an effective critic is established, this habit can be turned to examination of one's own work. Self-criticism is the best type of all because it can be applied in private and before effort has been wasted.

Finally, I must mention that building and making things is a particularly good type of training in strategic thinking, particularly if the thing built or made has some complexity. When children build model kits or make things out of Lego® bricks or using a construction set, they learn how to integrate many separate objects, how to convert a design into a finished object, and how to overcome obstacles. Home remodeling or building an addition has all the elements of a typical logistics problem, as well as being a puzzle solving exercise. Such experiences are invaluable, but are becoming increasingly rare for children due to urbanization and the elimination of technical courses in many schools.

Overall, there are many types of models, structures, and experiences that serve to exercise strategic thinking component skills or develop one's overall facility with this approach.

3

Scientific discovery in practice

The second chapter of this book presented the techniques of strategic thinking in terms of the nature of problem solving, how discoveries are made and tested, how creativity can be harnessed and enhanced, and how attitude and style affect strategic problem solving effectiveness. This provides the foundation for actual strategic problem solving in terms of the mastery of one's cognitive tools and abilities. Such an understanding is essential to success, but there is more to it than this. There are characteristics of the problems themselves that one must also get a feel for before even the best strategist can be successful. To make an analogy, it is not sufficient for the sculptor to have an aesthetic vision, to know the symbolism he wishes to convey, and to have harnessed his mental faculties to the task, it is also essential that he understand the nature of his materials: the way that different woods respond to carving and polishing, the receptivity of these woods to stain, the susceptibility of different types of stone to cracking, and so on. That is, one must know something about the subject to be strategized about per se. This is usually considered domain-specific knowledge (of cars or diseases, of pizza or electronics) which of course is beyond the scope of this book because there are hundreds of domains and their intersections where people are faced with problems. However, there are common elements across these different domains, common factors that may be leveraged into an understanding of the system such that one can effectively apply the strategic problem solving skills presented above. For example, we can note that any object or device or system behaves differently when scaled up to a larger size. This provides an angle for looking at problems. As another example, in any process there are potentially bottlenecks and constraints. Using the search for these properties as a guideline can increase one's problem solving potential.

To the strategist, then, a pile of sand, the stock market, and earthquakes may share certain similarities (all may exhibit cascades on all scales, following a similar power law relationship, and perhaps for similar reasons). A ship capsizing and a coup may share certain stability properties, and again for similar reasons. These are not just pretty analogies, but rather are structural analogies: cases where similar results obtain for similar causal reasons. That is, in these cases there are general rules of behavior that extend across domains. We may use these rules and general properties in strategic problem solving. The reason it is so useful to do so is that in real problem solving we are always faced with problems that have an element of novelty (for problems where this novelty is missing, we have merely business as usual and need not be clever at all). For novel problems we need to seek efficient ways of coming to an understanding of the problem or system because otherwise we can waste incredible amounts of time learning all about it (which often represents old conventional wisdom about it in any case). This section explores these general principles and ways of approaching and dissecting problems, using concrete examples and analyses.

Few of us can lift the front end of a car, but with a jack even a child can. This is leverage. The essence of strategic problem solving in practice is the use of leverage. With leverage, we can identify problems more effectively, expend less time and effort analyzing a problem, and exert more control over a system.

Leverage is a theme song or ad jingle so catchy that people whistle it. Leverage is a cruise missile that can fly right to a designated building and go down the ventilation pipe. We may contrast this to a brute force approach that uses dozens of high altitude bombers to hit the same target. Leverage is the polymerase chain reaction which can amplify trace amounts of DNA. The strategic thinker shuns brute force approaches because his or her time and energy are valuable and need to be used to obtain the highest return on invested effort. He or she seeks not to be busy but to be effective.

As mentioned in the first section of Chapter 2, the most important single task for the strategic problem solver is finding a problem, since in real life problems don't arrive as numbered homework assignments. Problems can be business opportunities, ideas for an essay, an idea for an invention, or "problems" in the negative sense of perhaps recognizing that one's company has no new products in the pipeline. In contrast to accepting the conventional wisdom about what problems exist or need a solution, the strategic thinker looks for leverage points: a problem no one has noticed, a place where a new technique can be applied, a controversy that can be resolved elegantly. Problems identified in this way are more likely to be solvable with less effort and therefore to have a higher payoff.

Once a problem has been identified, there are certain effective tools of analysis that can be applied. In contrast to brainstorming and lateral thinking, which mainly just help generate alternatives for simple (one-step) problems, the strategic thinker needs tools that allow him to reason through multistep problems. A number of such tools are presented in this chapter.

The objective of the strategic thinker is to generate solutions or products. To do so, he must apply leverage where it can be applied effectively so as to understand the system (in a research setting) or to control or manipulate it otherwise. For example, in science one of the key problem elements is determining what about the system may be reliably measured and which of these measures tells us something useful. For systems we wish to control (the profits of a business, the economy, a fish population) we must clearly identify what are the control points and how these determine system behavior. In this section these issues are dealt with.

3.1 PROBLEM FINDING

How does one find a problem worth working on? This is the most serious issue facing graduate students trying to start their research. Among practicing scientists, success at problem finding can make or break a career. In business, one may ask how one can recognize a new business opportunity or trend. Conversely, how can one recognize an organizational dysfunction?

The simple answer is that problems (including opportunities), just like diseases, have symptoms. An inconsistent theory will generate contradictions. An incomplete or inadequate theory will be confronted by anomalous data. Unexploited economic opportunities will generate unusual behaviors, dislocations, and unexpected rapid growth by certain companies. Undiscovered phenomena or processes will generate behaviors that can't be explained. In all of these cases we observe something that doesn't fit. In particular, we are looking for something that doesn't fit that is important or that has implications for other issues. This section explores these topics.

Paradox and contradiction

In searching for critical aspects of a complex problem, it is useful to develop an eye for paradox and contradiction. By focusing on a point of paradox, contradiction, or contention, one can avoid the need to learn everything about a topic. In any domain of knowledge, many things are well known to the experts, but to learn everything that the expert knows about a topic means that one must become an expert as well. This is clearly not efficient (or even possible) if many problems from diverse areas must be dealt with on a regular basis. In contrast, by focusing on a paradox or contradiction, one can locate those aspects of a complex problem that are not understood by others and that represent potentially key dimensions of the problem. In this sense, paradox indicates an opportunity, because it is a point of leverage. By focusing on a paradox, one can identify the key elements of knowledge and focus in on these, rather than trying to learn everything about a topic. The paradox also helps clarify what the problem is that needs to be solved. It is not the case, of course, that every paradox one uncovers indicates a problem one should solve, but they do indicate problems that may be attacked expediently.

Some examples help clarify the utility of paradox in identifying problems and opportunities. The existence of stagflation in the 1970s in the United States was a paradox: it was not supposed to be possible according to standard monetary theory to have both inflation and a recession together. Stagflation indicated an inadequacy in this theory. As a second example, the standard industrial theory over much of the last 100 years promoted economies of scale as a universal industrial law. The consistent result that medium-sized corporations outdo giants in terms of innovation (patents and new products per dollar spent), growth, and return on investment (Drucker, 1985) is a clear contradiction pointing to some type of diminishing returns with size due to internal wasted effort. If big corporations win, it is not due to efficiency, but is due to size per se. Size, for example, permits huge ad campaigns to be launched and national chains of dealers to be supported, but these are not "economies of scale." Rather, we may call this the bully factor. As another example, some entirely different approaches to psychological therapy (Gestalt, Adlerian, Rogerian, Rolfing, etc.) were all shown to be highly effective when first introduced. The paradox was that these different schools of psychology made entirely different assumptions about the mind and the emotions such that these different techniques could not possibly all be correct. It has turned out that the initial success of these techniques could in each case be attributed to the charisma of the founder. When attracted by a charismatic leader, people are much more motivated to change, and technique makes very little difference. This analysis is confirmed by the fact that the initial good results of the new therapies could not be replicated by later practitioners.

It is a safe bet that the conventional wisdom is often wrong. If it were not, then there would be no surprises, no massive layoffs, no business failures and no room for improvement and innovation. It is not the case, however, that the conventional wisdom is always wrong, so a pure contrarian policy is not useful and may be disastrous. Instead, contradictions indicate when and where conventional wisdom is wrong, and thus provide an opportunity to be ahead of the game. Several examples of domains where paradox can prove useful or informative are discussed next.

Controversy and paradox

A heated debate, especially if both sides are highly polarized, is an indication that all parties involved are missing a key to the problem, such that an opportunity exists for someone who can identify what they are missing. It is useful in such cases to ask "how could both sides in this debate be wrong?" Such polarized debates often result from different parties or schools of thought viewing a problem from a different perspective or scale (see next section), or with a different objective. The different points of view are often tacit, however, which is why such debates can continue indefinitely. A typical difference of this type occurs when one person is considering an issue at a strategic level (say, what should be the outcomes of schooling) and another is viewing it from a tactical level (which methods of instruction should be used). Not only does this cause confusion, but different strategic goals may require different tactical means.

Disconnects can occur when key words are used in different ways (e.g. economy, opportunity, rights, big government, environment). People generally assume that they are understood when they speak, but many "big" words (not those with many letters, but those full of meaning and connotation) can differ drastically in their usage. What is evident to the strategic thinker is that such different usages cause the same "facts" to lead to opposite conclusions, or similar statements to mean different things. To one person, "reducing big government" means cutting welfare and reducing paperwork for businesses, but to another it means reducing the military. Thus the strategic solution to a debate (and controversy) about such an issue may not be negotiation or compromise, but first the simple (!) step of clarifying what each party means (though they may not wish to be clear).

Divergent points of view can also result from different perspectives or scales, again leading to the paradox that different parties looking at the same situation can describe it in completely different ways. For instance, if in a small town a mall is proposed, the person who says it will be good for business is correct in the short term because construction money will stimulate the economy for a year or two. The person who says it will be bad for business may be taking a longer term view about the eventual death of the old downtown and the creation of abandoned buildings. Thus both people are correct in spite of making seemingly contradictory statements, because they are using different time frames for assessing outcomes.

Thus we can see that a "problem" often exists (disagreement, debate, controversy) that is at root the result of a paradox: that language is inherently ambiguous and many aspects of meaning are implicit, but we act as though we always understand what we hear and expect that others understand what we say. Attempting to solve such problems at face value is a waste of time. What must first be done is to resolve the implicit elements or disambiguate the language so that the actual issues can be laid out.

Paradox in business

In business, a paradox is often indicative of an opportunity or of a problem, or both simultaneously. Such a paradox is symptomatic of a disconnect between products and consumer needs, or between various portions of the economy. Let us take as an example the decline of mainframe computers and the rise of the personal computer. When the first personal computers were introduced in the late 1970s, such as the Apple, they were pitiful little toys compared with mainframes, with 8 K memories, not much capability, and a toy programming language, BASIC. One can see that when initially introduced they allowed people to use a computer who had no access to a mainframe. But why did such toys continue to grow in popularity and eventually nearly supplant mainframes? At each step of their evolution they were slower, had less powerful software, and had less memory than a mainframe. Many of the purchasers over this period had access, even unlimited access, to mainframes, but still bought these little machines. For the big computer manufacturers this was a problem (which they still have failed to grasp) and as a situation it is a paradox: how could an "inferior" technology take over from a "superior" one?

The simple answer is that from the perspective of the user the big machines were not so great because of the use of batch and time-share processing. The first computers were strictly batch and used punch cards. Each job was run using the whole computer, and when it finished the next job would run. A user would deliver their box of punch cards to a window and come back to pick up the output later, only to find a typo prevented it from running. In the 1970s time sharing became common and interactive terminals became available to allow editing of files stored on disk instead of cards. Many users were hooked up to a large central computer which would rapidly switch from user to user, giving each a small slice of computational time. This gave many people access to the big machines without having to wait for hours for cards to be processed. For the user doing simple typing, the result was usually satisfactory, but delays were frequent when the computer got overloaded. A delay of a few seconds doesn't sound bad, but it means that many words can be typed without anything appearing on the screen, which is annoying and causes the person to stop. This problem was usually exacerbated by tendencies to put too many users on the system. The system managers could point to very good average response times, but to the user the delays were real.

Much worse, however, was that the time sharing concept was implemented by long cable connections, usually phone lines. These lines usually had limited capacity (300 to 2400 baud). This meant that graphics and graphics-based programs (WYSIWYG word processors, spreadsheets) could not run over the network. In addition, refreshing the screens of hundreds of remote graphics images would strain even a large mainframe. To make a pie chart, one had to use a command language-based tool that would create a file that would be sent to a printer in another building. Output could be picked up two hours later. Errors were not evident until it was picked up. Repeating this process several times to get the desired graph could take several days. In contrast, on a personal computer one obtains a plot instantly and can change the options and labels many times and get a finished plot in minutes. The same holds for spreadsheets, CAD programs, fancy word processors, etc. Thus it did not matter to the user that the big machine was faster; in actual use the personal computer was faster because it did not have to be time shared and did not have a bottleneck in the transfer of data over phone lines. In particular, the personal computer was always available. Further, key applications were simply not available on the mainframes. On a mainframe, software would be purchased that many people were likely to use. If an individual was the only one wishing to use a piece of software, the central computing department would not purchase it but it would be too expensive for the individual user. For example, a compiler for a major programming language could cost $20 000 to $80 000 for a large mainframe system (in 1980s dollars). This constraint could be quite serious for users with various specialized needs. Thus a revolution in computing resulted from a paradox: an inferior technology produced better response times for the user, fewer bottlenecks, and new types of applications, all at a reasonable cost. After more than 25 years of this revolution, it is hard to call upper-end personal computers "inferior" any more, but they are still less sophisticated and less powerful than a mainframe, so the paradox remains and the trend continues.

Drucker (1985) focuses on paradox as a way of recognizing business opportunities and describes seven sources of innovative opportunity. First, the unexpected success or failure should be focused on because it indicates a missed opportunity or a change in the market or other conditions. He points out that unexpected success is not always highlighted for upper management since it is not a "problem" and is therefore often not followed up on. Second, opportunities arise from incongruities between expectations and performance (e.g. a company in a growing market that has trouble remaining profitable) or between market assumptions and market realities. Third, a type of incongruity is a missing link in a process or service. Fourth, a change in industry and market structures creates paradoxes and opportunities. Fifth, demographic changes can create economic mismatches and gaps. Finally, changes in perception and the generation of new knowledge create many discontinuities and thereby business opportunities.

Paradox in science

In science the role of paradox and contradiction is particularly notable. One use is as a form of proof or argument, particularly in mathematics and logic. If we are attempting to prove something and in the process we create a contradiction, then we have either made a mistake or our premise is wrong. This type of analysis has proven useful for testing expert systems such as those that diagnose disease. Another use is to reason by using opposites. Here we start by assuming the opposite of what we are trying to prove and reasoning from there. If we obtain a contradiction then this proves our original hypothesis. This is useful when it is difficult to analyze the situation we are interested in but its opposite may be simple. For example, it may be hard to prove there are no exceptions to a given generality, but easy to prove that if an exception occurs then contradictions arise.

Paradox in experimental domains is often encountered. In some cases it may merely mean that the experiments in question are just too ambiguous (Collins and Pinch, 1993), in which case the wise researcher avoids the topic or finds another way to study it. In other cases, however, a paradox indicates that something is fundamentally wrong. In recent years major paradoxes have arisen in certain fields. For example, the expected flux of neutrinos from the nuclear reactions in the sun has not been observed. Although the detection experiments are tricky (Collins and Pinch, 1993) no fault in the experiments has yet been found that would remove the paradox. For another example, the estimated age of the universe (as of 1995) is less that the estimated age of some nearby galaxies – a paradox if there ever was one. Note that in these cases various assumptions and calculations underlie each estimate, so there is room for resolution of the paradox.

A major paradox has recently been uncovered in attempts to trace human origins. Ancient human fossils have been found not only in Africa, but in Europe and Asia. Fully modern humans appear to have originated about 200 000 years ago and then spread out from Africa (the single origin hypothesis). Did they mingle with or displace preexistent populations in these other regions? Or did the populations in other regions evolve in the same direction and produce modern humans from these ancient roots (the multiregional hypothesis)? Evidence from mitochondrial DNA (inherited only from the mother) seems to show a single origin for all modern humans from a fairly small population, presumably in Africa. Evidence in support of the multiregional hypothesis is that morphological features dating back a million years in these other regions appear to have continued unbroken into modern populations. An example is the shovel shaped incisors of modern Chinese which appear to be similar to those of ancient Peking man. These two views can not both be right. If all human mitochondrial DNA has a common, recent origin, then regional populations can not have ancient roots. This is a clear paradox. There are several possible solutions to this puzzle. The first is that either the DNA studies are full of holes or the morphological studies are all wet (i.e. one group or the other in this debate is incompetent). Alternatively, something may be missing. One possible explanation is that the combination of mitochondrial DNA with genomic DNA can cause incompatibilities, such that only the mitochondrial DNA of one population persists while the genomic DNA may be a mixture of the DNA from both populations (Treisman, 1995). This clever solution would resolve both types of evidence, but it remains to be confirmed.

Darwin faced a paradox that threatened his whole theory of evolution. If all organisms descended from common ancestors, how did organisms get to remote oceanic islands? If they could not get out there then a paradox existed. The religious concept of separate creations would destroy evolution as an explanation. Darwin first noted that all animals on these islands could either fly or could survive long periods without food or water (e.g. snakes, lizards) and thus survive on floating mats of vegetation such as he had seen far from land on his voyages on the *Beagle*. He then did experiments which showed that many plant seeds could float for many weeks and then germinate when planted in soil. He thus conducted research specifically to remove this paradox.

The removal of paradox was also central to Einstein's studies of space and motion that resulted in his theories of relativity (special and general). His thought experiments showed the essential paradoxes that arise when one travels at the speed of light if it is assumed that time is invariant (e.g. if one moves away from a clock at the speed of light the clock appears to stop because light can not reach the observer from later times). He was then able to resolve these paradoxes by allowing the passage of time to be relative to the speed of the observer, which led to his theory of curved space–time.

In some cases, contradiction in science can boil down to the same kinds of linguistic tangles that bedevil normal discourse. This occurs when words are too loosely used and in particular when terms are not defined relative to how the object or process is to be quantified. This lack of operationalism is particularly rife in the social sciences, as discussed in Section 2.2 "Discovery as a process."

Summary

It is clear that paradox is a powerful tool for both problem finding and problem solving. A paradox occurs either when people are talking past each other or when there is some inconsistency in a domain of knowledge. The existence of paradox represents an opportunity, whether the domain is business, political debate, or research.

The interesting thing is that there is a paradox in how successful problem solvers handle paradox. The person with no tolerance for ambiguity does not tolerate paradox and does not make a good problem solver because they seek closure too soon, thus being likely to make up their minds in favor of one side or the other in a controversy rather than looking deeper and uncovering the root of the problem.

Conversely, they may simply ignore paradox and contradiction. On the other hand, someone who tolerates all ambiguity without being bothered at all does not see problems to solve, is not bothered by inconsistency, and thinks things fit together that don't. If you can read the tabloids without getting uncomfortable then you fit into this category. The true strategic thinker simultaneously can tolerate ambiguity, but finds certain contradictions so unacceptable that they become paradoxes which call out for resolution. The art in this lies in focusing on the paradoxes that matter, that have many implications and ramifications. This is the "art" in the Art of the Soluble (Medawar, 1967).

Perspective and scale

Looking for perspective

For grappling with complex problems, perspective is essential. Perspective enables one to disentangle oneself from a problem that may be too close to be seen clearly, the forest for the trees difficulty. Because things that are too close to us come to be taken for granted, it can be very difficult to conceive of them as being different. Perspective enables new conceptions.

I was visiting a state park south of Tallahassee in Florida. The central feature of the park was a clear spring, arising out of a sinkhole and forming the starting point for a river. The area of the spring was about 500 ft (152 m) across and the sinkhole in the middle about 100 ft across. Looking out from the bank, the view was really beautiful, and fish could be seen darting about in the shallows. There was a tower built at the edge of the spring so one could look down into the sinkhole. I climbed the tower, about 70 ft (21 m) high, and looked down. This was a literal change in perspective that was very revealing. Because of the source of the water in a limestone sinkhole, the water was extremely clear. This meant that from above every aspect of the bottom and every fish could be seen, in depth, in 3-D, with crystal clarity. This was beautiful in itself, but of particular interest was what it revealed about the fish. Around the mouth of the sinkhole, where the water was flowing out, three large gar (*Lepisosteus* sp.) hung motionless in the water, pointing inwards. They had divided up the sinkhole and were spread evenly around it, evidently waiting for disoriented fish to emerge from the spring. One would never be able to see this phenomenon from a boat, because one would be too close and because the boat would disturb the fish. Many problems have this characteristic, that they are only visible from an appropriate distance (sometimes literally). Another example of literal perspective involves archaeology. The search for ancient cities from the ground is often frustrating, especially in the jungle. From the air, however, networks of ancient roads in the deserts of Iraq, radiating like the spokes of a wheel from vanished cities, have been easy to spot. In Mexico, networks of irrigation ditches 1000 years old have been observed from the air, though from the ground one can not detect them.

In American business, it is assumed that the only way to deal with an economic downturn is to lay people off or encourage early retirements. It is held that layoffs are regrettable, but that there is no choice and business is business, after all. In this situation, no search for alternatives is made because it is not believed that there is any alternative. The cost of this policy, however, is very high. During layoffs, good people are lost who may have long experience in the business, those remaining are scared and demoralized, and early retirement programs can be very costly. More seriously, those remaining become reluctant to sacrifice for the company or to stick their necks out; why should they? When business picks up, staff are insufficient and opportunities may be lost. If more staff are hired, the cost of searching for, interviewing, and training them, and the cost resulting from the initial inefficiency of new staff can be very high. Overall, this is a very inefficient system. Further, it has a very high cost to the worker who is laid off who may use up his life savings, lose his house, become depressed, and suffer marital difficulties. When we examine the Japanese system, we see that they do not lay off staff except under the most extreme conditions, and yet they remain competitive. How is this possible? A large portion of the salary of the average Japanese is in the form of year-end bonuses. During a downturn, bonuses can be reduced across the board and no one need be laid off. The tradeoff is security of employment against certainty of wages. Further, during a downturn idle workers may be utilized to paint and modernize the factory, thus turning wasted time into an investment opportunity. We may contrast a third system, in place in parts of Europe, where major industries are owned by the government. In a recession layoffs are unacceptable, but so are wage reductions, so the government subsidizes the business. This may seem to be an ideal solution but it promotes inefficiency and debt. Once we see that other systems exist that differ from the one we are familiar with, then it becomes possible to imagine other variations and combinations.

There are other aspects of living in different parts of the world that similarly have major consequences for many aspects of how people act and do business, but that the people involved simply take

for granted. For example, in certain rural parts of Europe, land ownership has been tightly held within families for dozens of generations,
with each family holding onto their small plot. The consequence of
this is that there is very little movement or mingling of different
peoples, a situation Americans would have trouble imagining. A further consequence is that when historical events resulted in large land
holdings, these have tended to be retained such that certain families
maintain inherited wealth and position (related to the system of peer-
age still retained in England). In complete contrast, among many
American Indian tribes, land is largely held in joint ownership by the
tribe, which makes such inequalities in land control impossible. In all
of these cases it is only with the perspective gained by comparing
different regions that we can see the consequences of various land
ownership systems, whereas those within the system may take it
entirely for granted and be unable to imagine any alternative.

The proper focus or perspective is essential when looking at
business performance. From a very close up perspective, one may
examine departments within a corporation and see apparent efficiency
and bustle. From a slightly larger perspective, however, one may see
conflicts between departments that reduce overall effectiveness. For
example, salesmen might pitch a product in ways that conflict with
what the product can actually do, causing dissatisfied customers and
lost repeat business. The goal of the assembly plant to keep the lines
running may pile up inventory and increase carrying charges. From a
higher vantage yet, the company might be producing a product that is
doomed, such as a toy with a short popularity or large, gas guzzling
cars, in which case efficiency (the close up view) is really irrelevant.
Thus a proper strategic analysis of any organization depends on proper
identification of the level(s) at which problems exist. To do this one
must fight against the difficulty that most people tend to focus largely
at a single level. The president of the company who tends to think in
terms of markets and taxes and finance may completely miss the fact
that his company can not compete because it hires poorly educated
people, does not train them, and provides antiquated tools and equipment. The plant manager whose focus is on productivity may have
little concept of and little or no control over the choice of product to
be manufactured.

A classic case of perspective change results from the photographs
of Earth taken from space. From our human perspective, events appear
very local, and boundaries appear real. One can not cross a national
boundary without passing a big sign and showing a passport. From

space, however, there are no lines, no borders, and the artificiality of our boundaries appears evident. Further, and quite dramatically, we can see from space that the whole Earth shares the same atmosphere and the same oceans. This image of a natural Earth from space has been a significant factor in influencing many people to change their perspective on many issues regarding the environment.

This example illustrates that a key to gaining perspective is recognizing boundaries and seeing that boundaries are often arbitrarily drawn. It is often the case that boundaries around academic disciplines are historical accidents, for example, and do not necessarily correspond to the boundaries of real problems. American engineering students tend to specialize in either mechanical or electrical engineering, but Japanese students study mechano-electrical engineering, which is much more useful for problems in which mechanical and electrical components are combined, as in VCRs. Part of the difficulty facing large cities is that the jurisdictional boundaries of the city and its suburbs do not correspond to the larger system (the megacity) within which problems and potential solutions actually exist. In my neighborhood, the city, county, school district, park district, and congressional district all have different borders and each overlaps with multiple other entities. We often draw a boundary on a map and designate a certain area as a park or nature preserve, but many animals range over a far wider area than this and of course have little idea that a particular area might be safer than other areas.

Problems can often be solved by recognizing that a certain boundary is artificial and can be overcome. For example, the traditional segregation of product types into hardware, clothing, drugs and sundries, groceries, flowers, etc. represented a set of boundaries defining discrete types of stores. Recognizing that these boundaries are arbitrary has led to the mega grocery store and the mega retail store (e.g. Walmart, K-Mart) in which several types of stores are combined, which has the advantage to the consumer that several types of items can be purchased at one stop, thus saving time. It could be argued that a real solution to the race issue must involve the recognition that racial categories (boundaries) are also arbitrary, which becomes more evident as more people intermarry.

Viewpoint perspectives

A type of perspective that must always be considered in human systems is the perspective of different viewpoints. People pay lip service to different points of view, but what they usually mean is that they recognize that people have differing agendas or goals or wants. More than this, though, the life experiences and training of an individual may influence literally how they see the world. The city person standing by the river bank sees a beautiful spot and dreams of building a house there. The naturalist, on the other hand, observes the mud stains 12 ft (3.7 m) up on all the trees and notes that only trees that can tolerate prolonged flooding grow there, thereby concluding that the location floods for prolonged periods and at great depth. I knew an older couple in Colorado who had spent their entire life out on the flat plains, and could not tolerate going up in the mountains. They were used to a certain landscape. Upon living in the Pacific Northwest for several years, where the vegetation is lush and deep green, I was struck upon returning to the South with the signs of nutrient deficiency (yellowed foliage, particularly). My perspective had changed. To one person, an old neighborhood just looks old and they want a new house. To another, the large old homes have character and their low price suggests an opportunity for renovation at a profit. We can say that not merely do people react differently to what they see, but they actually see it differently. The recognition of how one's own perspective and that of others can be colored by viewpoint can help one avoid personal blind spots and biases and also help one understand the actions of others.

Perspectives in time

A change in perspective that can be crucial to understanding involves viewing events at different timescales. For example, barrier islands in the U.S. Southeast regularly erode at the southern end and build up at the northern end at rates that can exceed several feet per year in places. One should be very careful about where one puts a hotel under such conditions. As another example, the Mississippi Delta has lost huge amounts of land over the last 50 years due to subsidence because sediment from the Mississippi has been rerouted by levees. The Nile Delta is likewise diminishing because sediment has been retained by dams. In these examples, what we take to be fixed and permanent is clearly not so, even on a human timescale. The converse of this is also true: although we view the landscape as fixed and permanent, we don't believe that any of the changes that humans make (dams, farming, canals) are permanent, and yet the irrigation networks of the Maya can still be seen and early Bronze Age copper mines are still devoid of plant life in Britain and elsewhere.

When we take too short a time perspective, the result is often that we can not understand what is occurring. If we observe our yard for a while, we may see a bird flitting about. The motion seems random. If we observe for many hours we may observe the bird repeatedly returning to the same tree and discern that it is building a nest. Ants seem to wander at random, but with long observation we can note that when an ant returns from a location where there is a large food supply (a dropped banana or dead beetle) other ants begin to follow out the same trail and find the food also. With a short-term perspective, the opening of a new highway always seems to improve traffic. On a longer timescale, however, it happens that people begin to move farther from their jobs because of the easy commute. They start traveling farther to shop. The result is that traffic on the new highway increases even when the population does not. In all of these cases, short-term observations provide very little information.

A matter of scale

A related type of perspective is gained by considering issues of scale. We may consider scale in terms of physical size, speed, and complexity. In any situation, if one increases the scale (bigger, faster, more complex), the nature of the system response changes in a consistent manner, and often drastically. Scale must inherently be considered in all aspects of engineering, but also enters into managing an organization, writing software, and understanding natural phenomena.

The most obvious type of scale effect is that resulting from changes in size. A mouse and an elephant must inherently be built differently because an elephant with legs proportioned like a mouse would not be able to stand without breaking a bone. This is because bone strength is proportional to cross-sectional area, but stresses on the bone are proportional to length (due to bending forces) and because the mass of the elephant increases as a function of volume. Several other interesting consequences follow from changes in scale, without which elephants would not be successful. First, larger size makes elephants completely safe from predators (except humans). Large size also increases their life span because of physiological advantages that are not yet well understood. Of particular importance is that the large size of the gut of the elephant makes it possible for them to obtain nutrition from lower quality food because they are able to retain the food for longer and achieve more complete digestion. This is very fortunate because it would not be possible for an elephant to obtain sufficient high quality food (fresh leaves, fruits, seeds, etc.), nor could it nibble at such small items very effectively. Large animals must cover more ground in search of food and water, which makes communication more difficult, but again the elephant's large size provides an advantage. The large size of its vocal cords causes the sounds it produces to be very low in pitch and its larger ear bones are better able to hear such low sounds than ours are. Such low pitched sounds carry much farther, up to several miles, compared with high pitched sounds, which helps elephants stay in touch with each other. A further advantage of size is that a very large animal can have very large fat deposits which can help it survive hard times. Thus an elephant can easily go days without food, whereas a shrew, with its high metabolic rate, is in danger of starvation in a matter of hours. Thus the scale of the elephant produces a suite of new traits that cause it to live in a very different world from a mouse.

We may employ a formal analogy here and compare the elephant to the large corporation and the mouse to a small corner store. The metabolic intake of the corporation is the profit it makes on its products, and a large corporation can afford, just like the elephant, to subsist on low quality "food" (low profit margin items in high volume). The corporation has "fat" deposits (savings, cash, assets) that exceed that of the corner store and can afford to lose money for many months, whereas a sole proprietor will go bankrupt in a very short time. Like an elephant, a large corporation can not move very fast, but also like an elephant it is much less subject to predators. In the area of communication, the sole proprietor can call all his employees together for a meeting, whereas the corporation must employ slower, more formal means of communication such as memos, manuals, etc. Thus we can see many similarities that result from similar effects of physical scale.

Physical scale also affects all kinds of manufactured and engineered products, and as such must be considered by any inventor or manufacturer. As an object gets larger it will always be necessary for it to change its proportions or materials to meet the same goal. As a truck is increased from pickup size to semi-trailer size, the weight exceeds what tires can bear, and the number of axles must be increased. Larger airplanes require different proportions as well as different materials, including far stronger materials and connections, than a single passenger plane. In manufacturing, problems always crop up when scaling up from the bench top to full scale manufacturing. For example, it becomes very difficult to maintain constant conditions in a large vat where fermentation or some other microbial process is going on compared with a beaker size scale, and failures in large vats are quite common. A grass hut will stand up with almost any type of fastening, but larger structures must be connected in a much more rigid fashion. Thus issues of physical scale must be considered in any design or manufacturing process. In addition, consideration of scale can throw light on various types of failures (e.g. of bridges) and on the way in which larger entities (e.g. corporations) operate.

A second type of scale problem involves speed. Any time a process is speeded up, its nature changes in ways that must be taken into account. For example, the mechanism in a VCR must be more sophisticated than in an audio cassette player to prevent the tape from breaking at the higher speeds involved. In particular, when rewinding the video tape the VCR slows down as it approaches the end to prevent tape breakage. On a highway, exits must be long compared with the right angle turns of urban streets. High throughput systems such as overnight delivery services must operate very precisely and use sophisticated information systems. In general, the consequences of a glitch in a fast system are much worse than in a slow system. A person who stops their car to look at a street sign on a suburban street causes few problems, but stopping on the highway causes a crash. We may in general say that the higher the throughput (speed) of a system, the more likely that a glitch will cause turbulence. This is a direct analogy with flowing water in a stream, and applies to the flow of paperwork or to the flow of work on an assembly line. If you promise to fix a pair of shoes within a week it does not matter much if an employee is sick, but if you promise eyeglasses in an hour it matters very much. The drive for competitiveness in modern corporations seems to be sound, but as it pushes people to ever faster paces there becomes less and less room for human weakness. The person who is out sick for two weeks may be seen as being derelict and causing a horrible backlog. There is no room for the person with a good track record who is having a temporary depression (perhaps for good reasons such as grief). There is certainly no room for thinking, daydreaming, being creative, or learning new skills (see Section 2.1 "Strategic creativity"). In the long run, it is impossible to perform at a maximal rate month after month and year after year any more than one can sprint for as long as one can jog, and the consequence is the high rate of burnout currently observed among professionals and managers in large corporations. This is also why productivity often returns to previous levels some time after an efficiency expert increases productivity: the rapid pace of the "improved" system may be unsustainable.

Rather than cracking the whip and expecting people to work harder and faster for longer hours, true reengineering seeks out efficiencies in the system itself. Thus in flexible manufacturing (Bylinsky, 1994) the computer shoulders the burden of keeping track of special orders and parts so that workers can concentrate on the productive work of assembly. At Hallmark cards, the production cycle for new cards was shortened largely by bringing together the writers, artists, and others necessary to bring a card to completion rather than by terrorizing the employees with threats of layoffs. Thus the changes in system behavior with speed and the consequences of slowdowns in fast systems must always be considered in system design.

A further scale factor to consider results from the consequences of increases in complexity. An increase in complexity always increases some type of cost factor. In biology, increased complexity necessitates longer development times (longer gestation and immature stages in higher mammals, for example) and a significant expenditure of energy (the human brain uses a significant portion of the energy we expend). In an organization, complexity of function leads to an increased cost of coordination, commonly manifested in large numbers of meetings, memos, and phone calls. In computer software, complexity leads to an increased risk of bugs or of failure to complete the product. In computer hardware, high performance chips are very expensive to design and difficult to manufacture. From Section 2.3 "Strategic problem solving" we saw that this principle is general with all types of productions or systems, and in fact that failure risk goes up nonlinearly with complexity. This is why in the end the communist system in the Soviet Union failed: it is not possible to explicitly direct a complex economy from a central beauracracy. For this reason, self-correcting mechanisms are necessary. The complexity of modern society means that legislatures will always be in the mode of fixing things and tinkering with the laws and will never finish their job. The stability of the American governmental system results from its multiple systems for correcting errors (e.g. the judicial system will void a law that is uninterpretable or unenforceable). Complexly engineered products need many rounds of tinkering to get them right. Sophisticated scientific theories require a long period of debate and modification before they become finalized. Thus we can in general say that complexity is not free, and may in fact have a very high cost. The benefit of complexity, of course, is that it may enable far more powerful solutions to be achieved, but the potential cost of achieving such a solution should never be forgotten.

A perspective on perspective

We are creatures of habit, and tend to take our environment for granted. If our organization has a cumbersome and slow procurement department, we assume that it is not possible to do it better. If everyone around us is getting burned out, we get a little scared but assume it is inevitable. Without a little perspective, we do not see the big picture and even miss completely many things that are going on. Why do people build houses next to rivers that flood? Why do businesses act like they have never heard of economic cycles? It is a lack of perspective. Perspective is a key to understanding complex systems, to inventing, to scaling up, to design, to developing fruitful analogies, and to scientific investigations.

3.2   ANALYSIS: TOOLS OF THOUGHT

It is commonly known that success in any profession depends on both the possession of domain-specific knowledge (facts) and domain-specific techniques (tools). For example, a civil engineer must know about building materials, construction regulations, etc. as well as techniques for calculating buckling of columns and other factors. It is generally believed that there is an unbridgeable gulf between disciplines such that a person needs considerable retraining to change professions. While this is true for the domain-specific knowledge component of professional expertise, it is not entirely true for the techniques or tools of problem solving. Many tools of analysis are in fact quite similar across fields. A profession that shows this clearly is applied mathematics, whose practitioners apply their mathematical techniques across many different disciplines. Philosophers also may span many disciplines, but it is more difficult to verify that they do so successfully. Business leaders are similarly often able to successfully run several different types of businesses. In the arts, some individuals seem able to master multiple media (such as a singer–songwriter–dancer–actor).

The reason that some individuals can exhibit such versatility is that they can carry over their methods of reasoning, of problem solving, and of analysis from one field or problem to another. Even within a field, the creative and innovative professional generalizes across problems, whereas the plodder can only solve problems if they closely match problems whose solution has already been worked out. That is, the plodder does not apply much reasoning at all, but uses a cookbook approach. Thus, the successful problem solving professional, innovator, or scientist may be distinguished by the depth and breadth of their analysis or reasoning tools rather than just their mastery of domain knowledge or technical manipulative skills such as how to use a spreadsheet or oil paints.

The tools used by successful problem solvers are presented and illustrated in this section. These tools include evolutionary refinement, formal analogy, the detective model, classification, scenario generation, web and network analysis, failure analysis, and the use of mental constructs such as cycles and spirals. These tools are all used in multiple domains, though they are not necessarily all universal in their scope. These tools perform multiple roles in the overall problem solving process. In one sense they capture and formalize what were originally innovative reasoning processes. They thus represent higher order mental processes or in some ways an automation of complex reasoning steps. In another role, a creative step may involve the recognition that a certain problem may usefully be attacked by using a certain tool. Familiarity with these tools even affects one's perception of the Medawar Zone discussed earlier in this book, because skill in applying such tools puts otherwise unsolvable problems within reach.

Evolutionary refinement

Biological evolution is a totally blind process that nevertheless builds complexity and marvelous adaptations over time. It does so by a process of incremental change of existing structures combined with ruthless elimination of inadequate designs. A similar approach has utility in certain contexts as a problem solving strategy.

We may note several properties of the evolutionary process. First, at each stage in the evolution of a species the organisms must be functional. Second, changes are usually incremental. Third, early stages tend to be simpler. Software prototyping provides an example of this evolutionary approach. The formal approach to software development formally specifies the user requirements and the input and output requirements, and then uses these to formally develop the software structure and algorithms. After this is all done, coding begins. When coding is finished, the software is delivered to the user. However, it often occurs that the final program developed in this way is not really what the user wanted and a massive reworking must be undertaken. An alternative is prototyping. A simple, flexible version of what the user requested is developed. Following feedback from the user, it is improved and tested again in a continuous evolutionary fashion: it starts simple, functions at each stage, and is improved incrementally. Scientists also use evolutionary refinement. A research program will often start out with a preliminary or scoping study with a small sample size or crude measurements to test out methods, and then scale up to larger size or refined instrumentation.

This incremental approach is a very important strategy. A difficulty many people experience is that they think their entire creative product should be created all-of-a-piece. They think they should start the novel on page one or the software with formal specifications or the new product with a schematic. Such an approach is almost guaranteed to lead to writer's block or some type of other mental barrier. An evolutionary approach allows a crude problem solution to grow in refinement and complexity until it is adequate. We will see evolutionary refinement elements in many of the specific techniques presented next.

Formal analogy

Analogy is a significant problem solving strategy, but at the same time its importance is often overestimated. For instance, Koestler's (1964) theory of bisociation as the fundamental dynamic of the creative process makes overly much of analogy as a central tool of thought. His insights may be most applicable to humor where surprise caused by combining frames of reference is in fact the heart of the production process, and in modern art where bizarreness caused by merging conceptually remote images is also valued (consider the paintings of Salvador Dali). The combined image of a fish riding a bicycle can be amusing, but does not solve a problem.

It is useful first to clarify the distinction between poetic analogy and formal analogy. Poetic analogy (blowing autumn leaves are like refugees fleeing an oncoming army) is useful in the arts and is clearly related to creativity in such fields. In contrast, formal analogies are analogies in which the two things being compared actually share fundamental properties or behaviors in common such as turbulence in air and in water. Poetic analogies are only useful in real problem solving to the extent that they spur the memory to retrieve a tool, fact, or formal analogy that one can apply directly. Free association tends to generate mostly poetic analogies and is not terribly useful for practical problem solving.

Formal analogies are central to one's search through memory to retrieve a technique to solve known problem types (Langley and Jones,

1988). For example, many home repair problems involve glue or some other fastener (at least at my house). Analogy helps one retrieve from memory a case that is similar to the one currently being faced. In a calculus class, the key solution step is recognizing that a problem is similar to one that you have solved before. Once this is recognized, then one knows which types of techniques need to be applied. These examples involve first order analogies that refer one back to directly similar cases in prior experience. Even routine problem solvers use first order analogies on a regular basis. Giere (1994) argues that in science analogies of this type (problem A is similar to example B) are based on the actual similarity of problem type in terms of the model one would use to solve the problem (e.g. in physics problems). Thus such analogies are central to finding a method of solution.

We may next make a distinction between close and distant analogies. A close analogy is one where the things being compared have many properties in common. For example, an attacking army may be compared to a pack of wolves without stretching the comparison too much. More distant analogies are the product of primitive cognition (Arieti's paleologic [1976]). Paleologic analogies may share only one or a few features in common. Such a mode of thinking is common in dreams, in children, and in schizophrenics. When distant analogies are believed at face value, this is a symptom of mental illness, such as when a man concludes that he is Jesus because he is a carpenter or that the Moon is a big eye watching him because it is round like an eye. In the arts, distant analogies provide raw material for symbolic representations, and in this sense are directly useful, even though not taken literally. When they are played with, then they can become raw material for innovative thinking. This is because in problem solving we may need to emphasize only a single feature of an object or problem by which it resembles some other object or problem. For falling objects, color, shape, flavor, and texture become irrelevant; the objects are all "the same" (analogous) if their mass is the same. To find these types of similarities one must be willing to consider ways in which things are similar that differ from standard ways of viewing them. It is in this sense that De Bono's Lateral Thinking is useful; it allows distant analogies to be generated rapidly.

Distant analogies are central to innovative problem solving. Distant analogies often point to problem solving templates that one may be only loosely familiar with. For example, a sociologist might analogize that a revolutionary coup is like a boat flipping over. This analogy is more than poetic because both a government and a boat can exist in one of only two possible states but are unstable in intermediate states. Further, the transition between the two states is very rapid in both cases. This is a useful analogy for the sociologist, because the dynamics of boats are well characterized in engineering. Thus from common experience with boats he can recognize an analogy, and then go to the relevant literature (engineering in this case) and learn about formal models that might carry over to the problem of characterizing a coup.

This example also illustrates another benefit of distant analogies. The boat analogy is not perfect, but rather may provide a useful template or starting structure. In this sense, the distant analogy brings the problem solver into the proper domain of related models, within which his search can be directed more efficiently (Giere, 1994). Further work on the initial template may lead to the modification of various details to better solve the goal problem. Nevertheless, a fruitful distant analogy can get one on the right track.

Some other examples of the use of distant analogies further illustrate these points. Earthquakes are critical disaster generators, but are very difficult to study. One can not generate an earthquake experimentally, for example. Piles of sand onto which grains are gradually added produce sand avalanches that occur on all scales, from a few grains to a collapse of the pile. The distribution of sand avalanche sizes is similar to the distribution of earthquake sizes, and probably for similar reasons related to friction and the dissipation of energy. Thus the analogy between earthquakes and avalanches is more than poetic and suggests methods to study processes (such as earthquakes) that are otherwise hidden underground.

Thus distant formal analogies increase the breadth of models, tools, and techniques that the problem solver can bring to bear on a problem. That this has more than trivial implications can be seen from the well-established fact that outstandingly creative and productive scientists have particularly broad interests (even outside of science) compared with their more typical peers (Simonton, 1988). After being involved in making the atomic bomb during the war, Richard Feynman found himself unable to do research (because of mental fatigue from the war effort) until he began playing with models of spinning plates (experimentation performed in the cafeteria with plates spun on the end of his finger). This apparently silly activity was a productive distant analogy because it turned out that spinning plates provided a simple model of some serious phenomena at the atomic level (Feynman, 1984).

Everyone has observed tornadoes, traffic jams, static, and boats flipping over without necessarily having studied them in any detail. Such real world phenomena may provide useful models for similar processes that occur in business, design, engineering, physics, or any other realm of endeavor. A honeycomb provides a useful design starting place for a strong but light support structure and for optimal packing of objects. Radio static might provide useful ideas for information transmission failure or random walks on the stock market. Within one's specialty there are many case studies or examples that one has read about without having learned all the details. If an analogy can be recognized between the problem at hand and cases or phenomena that one is familiar with (but not an expert on), then one can look up the well-studied cases and use these as a starting point for further work. A facility with distant analogy as a method for quickly identifying relevant models obviates the need to know everything and thus makes the strategic thinker more efficient.

While analogy is a very powerful tool, and one that can produce what appear to be flashes of insight, it is not the only tool that successful strategic thinkers bring to bear. That is, a flash of analogy-insight is not the whole story. Other tools and techniques are presented next.

The detective model

For certain types of problems the best way to organize information and to solve the problem is to function as a detective. A detective must identify a list of possible suspects, and then compile information on each. A typical analysis looks like Table 3.1. This tabular analysis points to data needs and may even produce a match that implicates one of the suspects, thus solving the problem directly. One must always keep in mind, of course, that witnesses may be wrong, alibis can be faked, etc.

Table 3.1 *Detective model for problem solving, showing criteria versus suspects*

	Tom	Harvey	George
Motive	Y	N	Y
Alibi	N	N	Y
Opportunity	Y	Y	Y
Physical evidence	N	N	N
Witnesses	N	Y	Y
Criminal record	Y	Y	N

The detective model is the approach used by the U.S. Centers for Disease Control and Prevention to identify an unknown disease agent such as Legionnaires' disease or to trace the origin of a measles outbreak. In the Legionnaires' disease case, the disease was fairly quickly traced back to the hotel where the American Legionnaires had been having a convention, but isolation of the disease agent proved more difficult. Initial studies implicating smoking in cancer also used this approach.

A particularly interesting example of the detective model involves the tracking down of the causes of Kaposi's sarcoma (KS). This skin cancer is common in AIDS patients and in others with suppressed immune systems. It has been known that KS is common in parts of Africa, but the etiology of the disease was unclear. A detective-style approach helped clarify the nature of this disease. Initially, it was not clear if KS in Africa was due to a disease (such as HIV) or due to some other cause. Mapping the zones of frequent occurrence showed no overlap with HIV incidence, but there is an interesting overlap with regions dominated by volcanic soils (which was a very clever observation on the part of the person who made this connection). Further, KS in these areas differs from KS in AIDS patients. In AIDS, KS results from immune system failure and can occur anywhere on the body. In Africa, KS occurs largely on the lower legs and feet. This restricted occurrence tends to rule out a systemic disease agent (a conclusion confirmed by the lack of overlap with HIV), which presents a real puzzle. A final key clue in this detective story is that the individuals in Africa exhibiting KS are generally poor farmers who farm barefoot (wealthier farmers can afford shoes). Examination of the lymph nodes of these farmers shows that by walking barefoot they get small bits of volcanic pumice under their skin. These pumice bits are mostly trapped in the lymph nodes of the lower leg, but they are persistent and difficult for the body to eliminate. They lead to a local immune suppression which can be demonstrated by comparing white cell activity in the lower leg and in the rest of the body. Thus KS occurs in the immune-suppressed areas of the lower leg and feet. Thus the detective model, by successively eliminating suspects based on facts as they are accumulated, can solve even very puzzling problems. Note that this is also an evolutionary reasoning approach.

Another example of the detective model comes from closer to home. In fact, it involves homes: the search for one, to be precise. When searching for a home, one generally has a list of criteria: three bedrooms, family room, deck, not too old, fenced yard, etc. These are like the evidence in the criminal investigation, and houses one looks at are the suspects. Certain of these criteria are nonnegotiable, such as the number of bedrooms, and a house that fails to fit is eliminated. Others are somewhat soft, and are subject to some compromise. The approach one uses may also be very much like the techniques used to track down a disease. First, one might limit one's search to a given distance from work. Next, one might narrow down to an area where the schools are good. Next, one might identify areas close to shopping, but not too close. Then, one might begin actually inspecting houses for the detailed fit to the given criteria. When the list of suspects is exhausted (there are no homes under your desired price with three bedrooms that are near a good school) then one or more of the criteria must be loosened.

In all of these cases, the detective model involves progressively narrowing down the universe of possible suspects. In the case of Kaposi's sarcoma in Africa, this involved first narrowing down to a geographic area characterized by a particular type of soil, and then narrowing down to a class of people (poor farmers) who did not wear shoes. In the case of house hunting, regions of the city are progressively narrowed down and then individual houses looked at. Overall, this is a very common and effective problem solving technique.

Classification

For many problems, a very useful strategy (or at least a starting point) is classification (which includes decomposition). One may decompose a population into subpopulations or a process into subprocesses or even causal factors into component factors. We can also sort and lump in different ways. For example, when we break the human population into smokers versus nonsmokers, the lung cancer rate obviously differs. As another example, it is usually found that incidence of various diseases and causes of death differ between men and women.

The ability to make useful class distinctions is a central function of language and shows successive refinement with maturity (Giere, 1994). To a two-year old, all animals may be "doggies," whereas an adult recognizes hundreds of kinds of animals. People generally recognize more categories of objects in domains that are more important to them. A city slicker might not be able to distinguish different types of weeds or grasses but knows many types of flowers and house plants, whereas a farmer distinguishes not only a whole host of weed species but also recognizes varieties of corn and soybeans.

For strategic problem solving, the key point is that the solution to a problem may require a resorting into different classes. To the concrete-minded person, each object has a name, and that is that. But it may be useful to lump into the class "weeds" not just small plants like dandelions, but also fast food restaurants and traveling con artists. For some purposes we might lump all kitchen items but for other purposes all electric appliances, both kitchen and otherwise.

A particular type of classification is a hierarchical one, also called a tree diagram. Such a classification has many applications. Hierarchies actually exist in most human organizations, and are manifested as an organizational chart. Tree classifications are used for sorting out which species evolved from which and how they are related (how closely related, how long ago they separated). There are specific mathematical tools for deriving such trees from trait data. We may use a hierarchical classification to store information in a database, of which a library cataloging system is a classic example. We may conceive of a hierarchy of needs, as in psychology, or of tasks, as in planning. Well-structured computer programs have hierarchical features. Thus hierarchical structures have many uses in describing the real world and also for organizing our work, information, or plans.

When working with organizational hierarchies, it is best to remember that they only represent the lines of authority, but that work may flow or be regulated in an entirely different way. Different departments at the same level may put demands on each other (e.g. marketing, engineering, manufacturing) but not have any official authority to do so. This is a frequent source of poor organizational performance, particularly if disputes between such departments must go very high up the hierarchy to get resolved. Low-level individuals in accounting, safety, or procurement may have veto power or signoff authority over actions by individuals at all levels of an organization. It may sometimes occur that an individual can get work done by service components of an organization as though they were subordinate, even though they are not in an organizational sense. Thus there are actually two organization charts: the authority chart and the actual influence or control chart. It is the latter that one must learn in order to get things done.

The solution of real world problems frequently involves the decomposition of a process or function into its parts. For example, a dam in an arid region is only useful if it doesn't lose more water than it saves. Some Western dams were built with the assurance that evaporation would not be excessive and promptly lost millions of gallons into the ground because the lakes were constructed in valleys with sandstone walls. In this case all of the pathways of water loss were not taken into account (a full decomposition was not performed).

The recognition of new categories can resolve paradoxes and solve puzzles in many cases. For example, the terms liberal and conservative seem to have lost their meaning in American politics. Voters are constantly voting in puzzling ways and politicians often don't seem to fit with the party they nominally belong to. A key to this puzzle is that people's attitudes on economic and social issues can vary independently. Some people are conservative on both dimensions (the classic conservative who is against big government, against social spending, etc.), but others may be economic conservatives and social liberals or vice versa. Further, some individuals take an individualistic approach to issues, rather than sticking to a party line. Once these new categories are recognized, the modern political muddle makes more sense (though the media seem to prefer the old, and simpler, two category system).

Classification and decomposition also play a role in the arts. A person with musical skill can distinguish the parts played by different instruments. This is a type of decomposition that is critical to the ability to compose music. In most art, facial features are defined with respect to each other, and are thus not independent. One of the striking things about Picasso's work is that he decomposed facial features into separate units (eyes, nose, mouth, ears) which he treated independently with respect to position, angle, and even emotional tone.

The ability to classify, reclassify, and decompose are related to the use of analogy in a deep way. In usual classifications into, say, trees versus grasses, we are using some concrete trait (size, form) to classify. In analogy, we are using more distant or surprising traits to classify or notice a similarity. Thus if I say that the telephone is a kitchen appliance, it is not because we use it as a pounder to tenderize meat, and on first glance one might not count it as a kitchen appliance. In this analogy the connections that put the telephone into the class of kitchen appliances are less direct and one has to think to find them (e.g. it is used to order pizza, to call a friend for a recipe, to talk on while washing dishes to alleviate boredom, etc.). We can thus say that all analogies involve classifications, but that not all classifications (or decompositions) are analogical.

Classification is similarly at the core of the retrieval of memories in the sense that one often remembers classes as higher level entities or groupings, and that similarly classed things are easy to remember together (e.g. the typical barnyard animals can almost be pictured as a group). Memories are not, of course, filed away by classes like the card catalog in the library, but related things do have stronger associations or linkages in memory.

Some argue that mental operations are nothing but the recognition of and operations on classes. While at a very low level there may be some truth to this, it is like saying that music is nothing but notes; such a view does not help make one a better problem solver in the first case or a better musician in the second. What is particularly useful is to move beyond the concept that categories refer only to objects, such as furniture or foods, and realize that one also uses categories for processes. For example, most people have a concept of oscillation (springs, pendulums) (Giere, 1994), and would not have trouble extending such a concept to the stock market. Giere (1994) argues that hierarchical and radial structures exist in scientific theories (e.g. from the general concept and equation of a pendulum to more specialized cases of multiple connected springs). Realizing that one may classify not just objects but processes, one may explicitly search for models and categories that exhibit similarities, and solutions may be derived from this identification of similarity. For example, growth is a process-based category, of which there are various types, from simple to complex, but all of which share certain features in common. Oscillation, chaos, conflict, and communication are other examples. Thus classification is not merely about naming things, but may allow one to begin taking steps toward explicit modeling of a phenomenon.

Scenarios

Scenarios are a good problem solving tool that are particularly useful for planning. A scenario is an ordered sequence of events that either necessarily happen in the stated order, are hypothesized to happen in a certain order, or represent a goal sequence of events. An insect life cycle (egg to larvae to adult and back to egg) represents a necessary scenario. We must discover it but the insect can not deviate from it because the stages are developmentally linked. The supposedly regular cycle of the rise and fall of civilizations is a hypothetical scenario. A typical goal scenario is the plan a student makes when he goes to college: go off to school, major in something fun, graduate with honors, make a lot of money after graduation. Thus a scenario is not an arbitrary sequence of events but rather is one that has some causal linkage structure. A scenario has the property that it represents a plausible (if not necessarily inevitable) sequence of events. A scenario may also trace out the sequence of consequences resulting from a given action. A dream or goal per se, on the other hand, may consist only of the desired end state but lack a picture of the sequence of steps that could lead to the goal. For example, millions of teenagers dream of being rock stars or actors, but do not have any idea about the sequence of steps that would lead to their goal (and usually lack even the concept that steps toward the goal are necessary).

In order to imagine outcomes of our possible plans or to construct scenarios, it is necessary to be able to trace the effects of various actions or possible changes. A well-known example concerns the ramifying effects of the automobile on the family and on the city. The car allowed easier travel to work from a distance, and thereby created the suburbs. It also made dating easier and took teens away from parental eyes. It clearly had an effect on breaking up extended families, as well. Let us take a little less widely known example. Can the reader trace out the consequences for China of their language being picture-word based rather than alphabet based? Such a written language is surely harder to learn, since there is no relation between the symbol for a word and the sound of the word, but there are more serious consequences as well. A bigger problem is that such a language lacks an alphabetical order for ordering lists. Without this, it is impossible to have a phone book because there is no way to order the entries. No directory assistance. No rolodex. No card catalog at the library (not in Chinese, anyway). No database of customers. If you go to the hotel front desk and ask if Mr. Wu has checked in, they must look through the entire list of guests and find a name that *sounds* like "Wu." No typewriters; only recently have the Chinese developed menu-driven software with which they can select words to go into a document. How do you file personnel records or tax returns? This one change in language makes all the difference and creates ramifying consequences that impede the modernization of China. In contrast, Hebrew and Arabic, which also look strange to Europeans, are alphabet based and have no such problems.

Scenarios are very powerful when they are effectively applied. A business plan is a scenario, as is an ideal career path. Scenarios can also be used to trace out adverse consequences. For example, in a military campaign several different scenarios are usually worked out in advance to cover different circumstances, such as what action to take in the event of a reversal or of bad weather. During World War II, tracing out the scenarios resulting from different methods of countering Hitler's military led to the conclusion that air attacks against industrial targets such as ball-bearing plants would be the most effective. This scenario proved to be accurate, and a turning point was reached when German factories were unable to keep up the production of planes at the rate they were being destroyed, leading to total air superiority by the Allies.

A problem with the use of scenarios is that merely by being put in this form they may appear to be more certain than they are. Stock market forecasts, for example, are often phrased as though the prognosticator understands the causal factors affecting the market and can trace out their consequences over some months or years. In fact, most such forecasts are not much better than guesses. Scenarios generated by agencies such as the Central Intelligence Agency (CIA) to predict the consequences of certain world events or of U.S. policy are far less certain than they appear, and may bear no relation to reality in some cases. When the Soviet Union fell apart, no agency had such a scenario as even a remote possibility. Scenarios are often presented by political leaders to rally support for some policy or another, but such scenarios are often vague and not scientifically crafted. In fact, it is more likely that they are 80% wishful thinking or active deception.

Successful use of scenarios results from the confluence of several particular skills. First, the ability to order events in time is crucial (but is by no means universal, as can be seen by observing children and impulsive individuals). Second, the ability to trace out consequences of actions (if I study hard I will get good grades) is essential. Those tending toward magical or wishful thinking do not do well in this area. Third, the ability to envision alternatives is crucial. To reach a given goal played out by a scenario (e.g. go to college) it is useful to envision things that could interfere with this scenario so that they can be guarded against. Some imagination is helpful in this respect.

From this discussion we can see that plans and scenarios are related but are not equivalent. A goal scenario (go to college) can be the spur to development of a plan (save money, take college prep classes, etc.). A plan can include components to guard against adverse factors that may interfere with a goal scenario (e.g. the creation of backup plans for retreat in a battle). A plan can also lead to the generation of a sequence of steps, but these steps need not be related in the form of a scenario. For example, a plan to construct a factory includes finding a site, arranging financing, purchasing the land, designing the building, etc. Each of these steps does not logically cause the next, as in a battle scenario, and in fact many of the steps can be carried out in any order or in parallel. To take another example, a reorganization scenario is that a reorganization will lead to reduced costs and faster customer response which will lead to increased market share and higher stock prices. This is a (hypothetical) causal sequence. The plan to achieve the reorganization may involve many simultaneous actions as well as some sequential ones.

Given the above, it is possible to see strategic planning in a new light. Strategic planning is merely a subset of strategic thinking in general, which is why it has not been discussed until now. Strategic planning is usually goal directed, and as such usually has a goal envisioned as a scenario. A goal without a scenario is usually called a corporate "vision." For the reasons stated above, such "vision" things are pure fluff unless they lead to the generation of scenarios and plans to effect those scenarios. "Vision" statements of this type, floating high above the corporation like fluffy white clouds, actually increase cynicism in the ranks and should be avoided.

Strategic planning can be made much more effective by application of the techniques presented in this book. Many business motivational books have a lot to say about goals and motivation (in fact these are two of their main topics). If, however, one does not understand the system (markets, society, the engineering and manufacturability of your product, issues of efficiency and brittleness, feedbacks, bottlenecks, etc.), then goals can not necessarily be reached. The application of the techniques presented in this book should enable one to get a better understanding of the workings of complex systems so that one can move toward one's goals on the basis of knowledge rather than just enthusiasm. The ability to generate scenarios, including the possible role of adverse factors, is enhanced by the application of the combination of creative idea generation and structured problem solving. The verification of generated scenarios is made easier by the application of the various reality checks discussed in Section 2.4 "Reality check." The assessment of the overall feasibility of a strategic plan can benefit from the critical path model presented in Section 2.3 "Strategic problem solving." For example, a scenario that involves a long sequence of risky or difficult steps can be seen to be highly unlikely to succeed. If these risky steps can be reworked when errors come up (they are iterative type problems) then the overall risk goes down. An understanding of paradox, contradiction, bottlenecks, issues of scale, and other factors can assist one to recognize opportunities, construct scenarios, and make effective (rather than futile) plans to reach desired goals. Thus the techniques of strategic thinking not only provide assistance at various particular points of the strategic planning process, they also enable the big picture of strategic planning to be grasped and utilized more effectively.

Solution construction

As part of strategic planning, generation of scenarios, and other problem solving techniques, one must be able to generate a sequence of steps between where one is now and where one would like to be: this is the difference between the scenario and wishful thinking. Except for cases that are widely known (e.g. the steps involved in going to college), it is not always clear how to generate the step-by-step solution required. There are several specific techniques that may be used.

A very useful approach is to work backwards from the goal. To be a doctor, one must have a medical degree. To get a medical degree one must major in premed and get good grades and have a lot of money for tuition. Working backwards in this way the preconditions for each succeeding step can be determined.

Many problems are not so clearly defined as the go-to-college scenario. That is, we can not always lay out the full series of steps in detail. Consider taking a cross-country trip. It is easy to plan the general route and major areas to stop, but the details can not be determined from a distance. It is more effective to leave some flexibility in the plan because many details (roads under construction, tourist sites to see) can not be determined until one is close to a destination. Similarly, one can easily decide to go to college, but the exact courses to take must be contingent on availability, scheduling, success in prior courses, etc.

Sometimes working backwards fails because the obvious set of prerequisites to achieve a goal state are not possible. For example, the obvious requirement determined by working backwards from the goal of owning your own restaurant is to have $500 000 cash to buy one. The logical conclusion from this working backwards is to give up on the goal as unattainable. Another course of action is to see how one could construct a solution stepwise from the current condition. This might involve first getting a job as a restaurant manager to gain experience and credibility, meanwhile saving as much as possible. Then one might try purchasing a very small restaurant, such as a small deli. This could provide leverage to combine with some other business partners and a bank to secure the necessary financing to purchase the dream restaurant.

What if?

Related to the scenario is the What if? method of analysis. In the what if approach, you change a rule or law of operation and trace out the consequences. In the random rule change case, one just gets silly stuff that kids like to make up (What if pigs had wings? What if we lived on the Moon?). In a more sophisticated vein, impossible (or improbable) what ifs provide a basis for science fiction stories (e.g. What if we lived forever? What if some people could read minds? What if robots revolted?). Humor is profitably obtained by what if changes in the normal rules. What if a man got pregnant? has been the basis for several movies. What if one combines sports (e.g. luge and bowling)? has been the basis for some funny beer commercials. What ifs also provide a basis for serious problem solving, however. What if trains could fly? This sounds absurd, but is the basis for maglev (magnetic levitation) trains. The metal wheels of a train produce terrible friction, much worse than rubber tires on a road. This friction and consequent heating and wear on the wheels limits train speed, increases fuel costs, and causes high maintenance costs for track and wheels. A maglev train eliminates this friction. Once the cost of raising the train off the ground is considered, the additional cost of making it go faster is small. Thus the silly concept of flying trains actually has led to a product. One must ask an additional question about maglev trains: what if the power fails? This is not a trivial consideration, because then you are dropping a speeding train at least a small distance. Tests of maglev trains dropped on their rubber wheels at high speed have shown problems. Thus "what if" is valuable for trouble shooting as well.

Other innovations can be obtained from what ifs. One may ask what would happen to the laws of Euclid if parallel lines meet eventually instead of staying the same distance apart. The result of this change in the rules is a system of geometry (geodesy) where two lines that are parallel at the equator meet at the poles. By asking what happens if space itself is curved, one gets the curved space–time of Einstein. We saw above that pigs with wings is silly, but maybe bacteria that produce human insulin is not so silly.

What ifs have been instrumental in the development of new consumer products. Overnight delivery services were founded on a what if that at the time seemed ludicrous. What if you could take your phone with you? sounded silly in 1960, but today one may hear a phone ring from a purse or a briefcase almost anywhere. What if you could watch more than one channel at once on TV? is now available. What if you didn't have to wait for your hamburgers? Changing this one rule of restaurant operation (that cooking of food starts after the order is placed) created a whole new industry and saved people huge amounts of time.

While changing rules arbitrarily is only useful in the domains of humor and art, selective changes in the rules can lead to whole new sciences, technologies, and products. The key is to trace out carefully the implications of a change in the rules. In social engineering, legislators are often playing what if, but their basis for making projections is rather fuzzy. It may be asserted that reducing capital gains taxation will increase investments, but we do not know for sure. We can't make a prototype. Three strikes and you're out prison sentencing similarly is based on an assertion of the effect of changing the rules, but without any firm guarantee. We often only find out "what if" long after such a change is made.

A subset of what if analysis is the playing of the devil's advocate. There is a strong tendency when doing any planning to assume that the best outcome will occur. In a business plan this temptation is quite strong, because otherwise investors might not be interested. It is very important to counter this tendency with a series of negative what ifs. We should always ask what if there is a disaster? This would point out the need for fire insurance, for a backup line of credit to get the business open quickly again, etc. We should ask what if the product we plan to make turns out to be a stinker? What if the economy hits the skids? What if I try medical school but find out I faint at the sight of blood? This line of reasoning leads us logically to the technique of failure analysis.

Failure analysis

It might seem paradoxical that a way to achieve success is to look for failure, but such is indeed the case. It is characteristic of a successful technical product that it doesn't fall down, blow up, catch fire, turn out to be incoherent, short out, or lead to lawsuits. A considerable portion of the expertise of the civil engineer is dedicated to verifying that his structures will remain standing. Such verification is achieved by understanding possible failure modes and demonstrating that they have been reasonably guarded against. For example, earthquakes exert several types of stresses on buildings, including shaking and swaying of the upper stories. Observations of modes of building failure have led to specific construction practices to prevent these types of failure.

In general, the solution to a problem or the creation of any device, product, plan or system should include as an integral part the analysis of possible modes of failure, their likelihood, and possible means of overcoming or preventing such failures. Volvo is famous for its investigation of accidents and the consequent design of impact-resistant cars. In Japan, large sums have been spent to earthquake-proof buildings. In battle, possible modes of failure include a break in the line, insufficient munitions, being circled and attacked from the rear, and panic among the troops. These possible modes of failure are guarded against, respectively, by secondary lines of defense, good logistical operations, tactics and intelligence, and troop discipline and training. No army has long survived that ignored modes of failure. In science the "failure" of a research project can come about due to insufficient sample size, uncontrolled variables, misunderstanding of basic theory, sloppy methods, poor record keeping, etc.

It is useful to distinguish different modes of failure. A *brittle* system is one that fails abruptly outside of its range of tolerance. For example, a tree snaps off when the wind exceeds a certain level. A *robust* system, in contrast, can continue to function under a range of conditions. For example, the B17 "flying fortress" was able to return from missions in spite of being all shot up and having pieces of wings missing. It is questionable whether modern high performance jets are robust in this sense. In a software system, graphical user interfaces and error trapping can enhance robustness.

Another important feature is whether failure is incremental (graceful) or catastrophic. An army that is losing a battle should be able to retreat in an orderly way. This characterized the Roman legions, for example, compared with many of the armies it fought which lacked discipline and would scatter if bested or if their leader was killed. In fact, they had a specific, highly ordered defense, the turtle, that they would implement if outnumbered or losing. With this defense the men closed ranks in small groups and presented a wall of shields with protruding spears. In this defensive posture they could regroup or retreat. A characteristic of modern economies compared with less developed economies is that modern economies have many mechanisms that make economic downturns more gradual and less catastrophic (insured savings accounts, lines of credit, unemployment insurance, etc.).

Several factors can contribute to robustness or grace under pressure. Distributed information systems can allow for incremental failure modes. For example, in World War II, American soldiers were briefed about a mission as an entire unit. This meant that if officers were killed (as was common on D-Day), the next ranking soldier could take over, even if conditions changed. In contrast, the German and Japanese militaries were much more hierarchical, and only officers were trusted with mission information. When experiencing success, their armies were highly effective, but when pressed by Allied attack and under shifting battle conditions, their system caused heavy casualties and contributed to their defeat.

Another factor promoting robustness is feedback in the system. While this is discussed in more detail in Section 3.3 in "Feedback and information," it is discussed briefly here. Feedback can provide information that leads to an adjustment in response to a change in conditions. Feedback that is weak or delayed can lead to a response that is too late, which may be catastrophic. In the American Civil War, the South's missing cavalry at the battle of Gettysburg is an example of this. Lee had intended to withdraw from engagement unless the ground was of his own choosing, but at the crucial moment when his army encountered the Union troops, his cavalry was nowhere in sight, so he could not tell if he was facing the entire Union army or just a small force. Lags in feedback can contribute to industrial cycles when businesses expand without realizing that market saturation has already occurred. Feedback that is timely can help prevent overshoot behavior. Effective feedback has been crucial in reducing inventory swings and over-ordering, which thereby reduces inventory carrying costs.

Analysis of failure should extend to consumer products. Such an analysis would both lead to well-designed products and allow anticipation of likely consumer complaints. The old AT&T telephone was well designed with respect to a common likely failure mode: being dropped. In contrast, one is well advised not to drop many modern phones. Another consumer example is hair dryers. A likely failure mode for hair dryers involves dropping them into water while turned on. This type of event not only destroys the hair dryer, but can kill. As a consequence (not necessarily of forward-looking failure analysis but more likely due to lawsuits) some of the better models are beginning to feature ground fault interrupt circuits (though one is still advised to keep them out of the bath). A common failure mode for a toaster is for toast to get stuck, which can cause a fire in a poorly designed model. In the mad dash to get new consumer products to market, there is a decreased tendency to check for modes of failure, even though doing so can increase consumer satisfaction, reduce returns, and reduce liability exposure.

Finally, it is noteworthy that an understanding of failure may directly solve a problem. For example, the key to understanding the operation of the AIDS virus (and consequent symptomology) was figuring out how it could cause the immune system to fail completely. Many historians have pondered the fall of the Roman Empire, which always seemed to have tremendous significance. This mega failure has been blamed on internal corruption and other political factors, but may have been largely the result of climatic changes (drying) in North Africa and the Mediterranean region which put pressure on the Romans' agricultural production, along with huge increases in the populations of surrounding regions such as northern Europe.

One may even observe a role for failure in the arts. It is often the case that a new musical or artistic style is a reaction to a perceived failure or deficiency in an existing style. Perceptions that realism is incapable of representing certain emotions or concepts, and that it was too confined in terms of the universe of possible visual patterns contributed to the rise of modern art. Physical failure of certain arts media to persist over time due to flaking, fading, sagging canvas, etc. has influenced how artists paint. Perceived failures in public buildings (e.g. spaces not to human scale, funneling of winds into passageways by structures, poor use of light) have influenced design in many ways.

Overall, failure analysis is an integral part of problem solving, while being at the same time a specific tool for organizing information. Failure analysis can be a useful type of reality check, as well, by helping one to take off the rose colored glasses and look specifically for what might make a product or system fail. The inventor often begins with a perception of where a product fails, and then seeks to alleviate this failure. Political opponents are motivated to probe for policy failures (though not necessarily to treat them honestly). Thus failure analysis has multiple uses and implications in strategic problem solving.

Cycles and spirals

A useful structure for organizing information or events, particularly events that occur over time, is to use some variation on the cycle, including the spiral. In a business like farming that is inherently tied in to the annual cycle of the seasons, planning, including financial planning, is based on this cycle. Many other cycles exist that can help one organize information and anticipate future changes. In business, it is widely known that there are underlying cycles of prosperity and recession. It is even known that some industries are more susceptible to such cycles (e.g. steel, autos, paper). In spite of this well-known cycle, banks commonly loan money as though downturns can not occur. Many business plans assume either that current conditions will continue or that things will trend upward. Far fewer business failures would occur (particularly in banking) if planning were based on a recognition that the economy is cyclic. The mega-bank and securities failure of late 2008 is a perfect example of this problem.

Another natural and well-known cycle is that of the establishment, aging, and renewal of neighborhoods. As homes age, they naturally require more upkeep. Expensive homes are usually not only better built, but are more worth maintaining. Lower income homes are more likely to decay into slum status. A complete cycle involves renewal of such decayed areas, which may occur by yuppiefication, bulldozing, or conversion to industrial or highway sites. In ancient times, by contrast, fire frequently performed this renewal function. Rome, for example, burned on a fairly regular basis (Boorstin, 1992). In other ancient cities, it was easy to tear down old houses because of their small size and simple construction. Thus the problem of slums in modern times may be at least partly a problem resulting from the higher cost of tearing down old housing.

It is very useful to think and plan in terms of cycles in many other areas as well. The growth of infants is loosely cyclic, and therefore so are their appetites (which drives parents to distraction). Manic depression is cyclic over months or years. An individual's level of alertness varies in a cyclic way over the day. Awareness of such cycles can assist in scheduling tasks for the most appropriate part of the day. Tasks such as jogging and doing arithmetic all have optimal times. For many people, the natural cycle includes a nap after lunch which is therefore a bad time to schedule a class if one is a student or to arrange a seminar in which the lights will be turned down. Coffee is often used to fight these natural physiological cycles, but it is perhaps better to take them into account and ride them to enhance productivity (i.e. do intellectual tasks when most alert, run errands and make phone calls when groggy) rather than to fight them.

The spiral is a thought construct that is also useful. We may visualize the spiral as like the swirl of water running down the drain. The spiral is commonly used to describe the course of alcoholism, and the down-the-drain quality is explicitly carried over. In this context, the spiraling (circular motion) represents the fact that the alcoholic is usually not uniformly drunk, but rather tends to cycle in and out of severe episodes. The downward direction represents the self-amplifying destructive trend overlaid on this cycle. During drunken episodes, the alcoholic misses work, which creates stress at work and at home, alienates family and friends, and damages his health. As these problems accrue, they amplify worry and depression which lead to further drinking. The spiral narrows as drinking becomes less episodic and more continuous. The alcoholic spiral really does resemble water running down a drain. This is a classic case of runaway positive feedback, which is discussed later in more detail.

A spiral structure may also be discerned in the process of economic development. On top of the ups and downs of a developing economy (the loops in the spiral), successful development leads to increases in productivity, infrastructure, and education that lift overall levels of wealth. There is also a tendency for boom and bust business cycles to become less pronounced over time. This structure is thus like an upside down bathtub spiral. Spiral structures may also be usefully applied to describing the maturation of a good marriage and other phenomena. Overall, cycles and spirals provide useful constructs or templates for understanding phenomena and for problem solving.

Transformations

A useful mental construct is the transformation. A transformation converts one type of thing into another type. For example, raw materials (A) enter a factory and are transformed into finished products (B), symbolized by the following:

$$A \rightarrow B$$

Upon marriage, a single person is transformed into a married person. Upon purchase, a new house becomes a home. A transformation is generally a discrete change in state and represents a qualitatively different condition. A person can become single again by divorce, but will never be the same in status or in behavior as a never-married person. Note that a second transformation is required (divorce) to undo the state attained by marriage. To convert a bottle back into a raw material, a transformation via grinding is necessary.

Transformations are useful for recognizing how one enters different states. In many cultures, the transformation from youth to adult has been heralded by a rite or ceremony. This was a discrete transformation, with the new adult being expected to act as an adult in distinct ways. In contrast, in modern cultures this transformation tends to be very gradual, with obtaining a driver's license and graduating from school being two of the few milestones that are widely recognized. In the absence of a clear and discrete transformation step, many young people in fact have difficulty making the transition to adulthood, and may continue to exhibit juvenile traits (irresponsibility, lack of romantic commitment, financial dependency) well into their late twenties.

Transformation is also the key to reaching goals. If it is decided that a matrix organization (one with multiple cross-departmental lines of communication and teaming) will be better than the current hierarchical organization, it is not sufficient to proclaim that this is the goal. One must design a transformation that will convert the current organization into the desired one. Failure to specify an efficient transformation will lead to chaos and low morale, as well as reduced productivity, and even to the eventual abandonment of the goal. It is often the case that it may even be easier to start from scratch than to transform an existing organization into a new form.

Let us consider transformations involving private versus state ownership of enterprises. When a nation decides to nationalize an industry such as railroads, it generally purchases it and makes it part of the government. This transformation (which may or may not be a good idea) is in some cases easily reversed by merely putting the enterprise up for sale if the enterprise has been run as a fairly independent entity and not overly beauracratized. If, however, the state owns almost everything, as in the former Soviet Union, then the reverse transformation is very difficult because the state owns most of the capital; there is no large pool of money available for making purchases of the enterprises for sale, nor businesses to act as agents or purchasers. That is, the bulk of assets are not liquid and are in fact owned by the seller. Thus the first step in such a transformation must be the distribution of state owned assets of ownership to make them liquid, which is what happened in Poland in the 1980s.

We generally think of transformations as reversible, though perhaps with difficulty (marriage is reversed by divorce), but there are many domains where this is not so. For example, the transformation called extinction is irreversible. Once a high crime slum is created, it is very difficult to undo. Once pollution is widely dispersed (e.g. by Chernobyl) it is impossible to retrieve.

It is useful to recognize transformations because following a transformation a qualitatively different structure is created. That is, a transformation does not just result in an incremental, additive change but in a structural change. A condemned house has an abruptly different utility. A prison sentence produces an irreversible change in one's future opportunities. A corporation must behave very differently from a sole proprietorship. Being deep in debt is a qualitatively different state in which one is in danger of losing one's house. The utility of the transformation model is also in the recognition of the nature of the transformations that produce a given change. Understanding transformations can enable one to utilize these transformations for achieving one's goals, or to avoid the occurrence of undesirable transformations.

Nets and webs

When we are faced with multiple factors in a problem context, it is particularly important that we possess tools for relating these factors. In the absence of such tools, there is a tendency to focus on single factors (Stanovich, 1992). For example, slums may be blamed on prejudice alone, when in fact slums have existed (and do exist) in racially uniform societies. On the other hand, merely making a list of factors that are related to a given problem is more likely to cause frustration than to lead quickly to a solution. Network and web diagrams and constructs provide powerful tools for dealing with such complex systems. We may consider as a simple case a spoke diagram. In this diagram, the item being affected is at the center and the factors impinging on it radiate out as spokes. This might apply to the factors impinging on an individual's health, each being independent and focusing on the individual.

When there are multiple entities involved, a web diagram can be used. In this diagram, the different entities or variables are connected by arrows showing which affect which. This can be used to illustrate the multiple interactions in an economy or social system. A similar diagram, a loop analysis, focuses on the elucidation of feedback loops (discussed in Section 3.3 in "Feedback and information"). For instance, if economic development enhances educational opportunities in a country, which enhances economic development, this would be a positive feedback loop. Economic development could simultaneously reduce the birth rate (a negative feedback loop) which would increase the chances of success for each child (a positive feedback), and so on. Each of these factors can be diagramed with arrows with pluses and minuses to show their effects and interactions. Such a diagram is a powerful tool for understanding the causal structure of a system. In the face of multiple loops, it is often the case that the timing and magnitude of the feedbacks determines the ultimate level and relative prominence of the different loops. In such a case, it may be necessary to resort to explicit calculations to determine the outcome, often using simulation. Computer simulation is a powerful tool for analyzing such complex interactions, although care must be taken because of the simplifying assumptions that are often introduced.

A workman and his tools: summary

One of the chief characteristics of an experienced craftsman is that he knows his tools, what they are capable of, when to use them, and how to get maximum performance from them. While no one doubts this for hand tools or for the artist, the role of tools in conceptual work seems completely overlooked. In school, only subject-specific analysis tools (how to do a return on investment calculation) are taught. Of far more importance for strategic thinking is the mastery of general tools of thought and analysis. In this chapter general constructs have been discussed for organizing information, for tracing causation, for classifying, and for generating analogies. In the following parts in this chapter, further tools and techniques are described in the context of general principles of system organization.

3.3   SOLUTIONS: GETTING A GRIP

If, in a rural corner of the world, we observe a small boy leading a large ox or water buffalo, we may wonder at his fearlessness. Closer examination, however, may reveal a rope passing through a ring in the animal's nose. This is an example of getting a grip on the problem (in this case, literally). The boy has found an effective control point for the system.

The strategic problem solver also seeks to get a grip on his system, either to learn more about or to manipulate and control it. It is not very useful to just know trivia about a system if you can't use these facts to some end. On the contrary, if you understand the key aspects of a problem such that you can solve the puzzle or manipulate the system, then the details are perhaps not so important. There are certain key dimensions of system behavior that specifically point to control points. For example, a tradeoff points to a constraint and is inherently a control point for determining outcomes. A bottleneck is a control point governing throughput and is a critical point for system control. Information flow can also be manipulated to control a system. The feedback structure of a system can point to effective control strategies as well as illuminating fruitless ones. This section addresses these aspects of the overall process of problem finding and problem analysis, discussed above, and problem solution.

Constraints and tradeoffs

A very effective strategy for controlling a system or solving a problem is to look for constraints. Constraints feature prominently in any design problem, such as inventing a new device or improving a product such as a computer. They are also central to understanding how things work, whether it is an economic system, a medical problem, or a social dysfunction. Tradeoffs are the usual consequence of constraints. A tradeoff occurs when we can have more of something only by giving up something else. Tradeoffs are not always obvious, and in their personal lives people often fool themselves that they can have it all, that there are no tradeoffs. An understanding of the constraints operative in any given setting provides a powerful tool of analysis. It can prevent one from wasting time trying to create something that can not exist or to do something that is impossible. For complex problems, the constraints operative in a system can provide real insight into how the system works, where it has weak points, and even may explain its dynamic behaviors.

The way in which constraints are typically overcome is either with information or with new technology. Information, by allowing us to organize, track, and sequence objects and actions, can help us overcome many constraints. Information itself, of course, has a cost, and can therefore also act as a constraint, a factor we will also explore. New technology may incorporate information, such as in computer controlled fuel injection in cars, or may result from the substitution of new materials or processes (which of course is also the result of new information produced by experimentation). In the sections that follow, the role of constraints in different domains is explored in the context of the roles of information and new technology in overcoming these constraints. This interplay applies generally to all types of innovation, discovery, and problem solving.

Technological constraints and tradeoffs

Nothing illustrates the role of constraints and tradeoffs as much as technical innovation in industrial design. We observe repeatedly that improvement of some product or device continues until some constraint is reached that causes further innovation to become increasingly difficult, at which point a new technology is needed to allow further improvements. New technology is inherently also about information. It is useful to explore several examples here.

When it became clear in the late 1970s that the days of cheap gasoline were over, there was a push toward (and legislation mandating) better fuel economy for cars. For many people, smaller cars provided a quick answer, but many vehicles must be larger to fulfill their purpose, such as family cars, utility trucks, pickups, vans, limousines, etc. For these vehicles, high gas prices were deadly. The process of change in the quest for better gas mileage is instructive for what it shows about constraints and technology. At the time, some auto industry representatives claimed that not much improvement in gas mileage was possible because of limits in the energy contained in gasoline, but it has turned out that a whole series of improvements (DeCicco and Ross, 1994) has led to substantial improvement. One of the most obvious ways to improve gas mileage was to reduce the weight of the car (besides just building a small car). Older models featured heavy frames, thick sheet metal exteriors, huge engines, and heavy interior construction. A considerable improvement in gas mileage was realized by reducing the excessive robustness of the body and frame. After a certain point, however, a constraint was reached because the body became too fragile to support ordinary use and to protect the passenger. Other sources of savings were necessary. The use of novel lightweight materials for bumpers represented an application of new technology that produced substantial weight reductions. Improvements in gas mileage at the point of combustion were also sought. Traditional carburetors did not allow for much improvement in mileage, but a new technology based on information, the computer controlled ignition system, did lead to substantial savings. Thus the modern car incorporates several information processing systems and high-tech materials. To obtain further savings, it was clear that engine weight needed to be reduced, but the use of aluminum blocks was not entirely satisfactory due to their tendency to warp at high temperatures. A potential solution to this that is currently being explored is the use of ceramic and ceramic/metal combinations for the engine, which represents the use of a novel material. It has also been found that new types of tires improve gas mileage as well as lasting longer. The combined result of all of these innovations is cars that routinely get over 30 mpg (9.4 litres/100 km), and sometimes over 40 mpg, figures that were regarded as impossible two decades ago. In the process of achieving these results, a number of constraints due to materials or technology had to be overcome by substituting materials, by introducing new technologies, and by utilizing information processing capabilities (in the ignition system). This entire process of innovation hinged on recognizing successive constraints and realizing that they could be overcome.

Computer technology provides another excellent example of the problem of constraints. For certain types of problems, speed is essential. For example, it will not suffice to take three weeks to run a computer algorithm to predict tomorrow's weather. Straightforward improvements in computers have led to a steady increase in central processor speed, but for the truly large problems special purpose computers have traditionally been used, the so-called supercomputers such as those at one time produced by Cray Research in Minnesota. These computers are very sophisticated and complicated. In particular, the central processor (CPU), which does the actual number crunching, is very complicated, which makes its design very difficult and expensive. Manufacturing such complicated devices has also been expensive, with quality control being critical. In spite of improvements in this technology with each passing year, the expense did not make these machines accessible to most users. There were in fact constraints in this problem that most people did not recognize. The result of these constraints was that foreseeable increases in CPU speed did not seem likely to be adequate to the new demands being put on computers. A completely new technology was therefore developed that did not depend on super fast CPUs: the massively parallel computer. It was noticed that most of the problems for which supercomputers were being used involved the repetition of the same computations for multiple sets of data. For example, in weather forecasting, the same computations are performed at thousands of points in space, and then repeated at the next time interval. For fluid dynamics problems, a grid of points defines where the flow of fluid must be computed. In parallel computing, many of these spatial points are fed to a grid of CPUs simultaneously and the set of computations is performed on them all at once (up to hundreds or thousands at a time). In this type of computer, the individual CPUs need not be complicated, and in fact need not even be blindingly fast. Their sheer numbers mean that even reasonably fast processors (which can be manufactured quite cheaply) can perform computations on large data sets at a speed comparable to a supercomputer, but at an overall cost that is much less. Clearly, in this case the constraints due to the bottleneck at the CPU were much more efficiently solved by using a totally different approach than by simply pushing the limits of silicon and miniaturization.

Biological constraints and tradeoffs

In biology, constraints are central to the design of organisms, although this fact has taken many years to uncover. Because organisms can not change their shape at will, there are necessary tradeoffs that cause an organism that is good at solving one problem to be bad at other problems. For example, the hollow stem of many weeds allows them to grow very fast (cheap construction), but it does not permit them to become the size of trees because of insufficient strength. The chitinous external skeleton of insects provides strength and protection, but at larger sizes (such as the size of a horse) the weight would be prohibitive, transmission of oxygen via diffusion would not be adequate, the internal attachment of muscles would be problematic, etc. Thus the giant organisms so popular in horror movies, such as giant spiders, can be clearly seen to be ludicrous because they violate some physical constraint.

For other examples, we can examine the form an animal takes. The optimal foot for running is either padded or hoofed. Such a foot can not be used for grasping and is not very good for digging or swimming. To be efficient at digesting grass, ruminants must have very large digestive systems, which limits how small they can be. The physical form of an animal constrains what it is good at, where it can live, and how it can obtain food.

Humans must endure such constraints as well. For example, in track and field competition, it has been found that the same individual can only rarely be competitive in both sprinting and distance events. This results from the fact that there are two types of muscle fiber, long and short. The short fibers have a quick response and are good for explosive action such as sprinting or boxing punches but they tire quickly. The long fibers can not respond as quickly but have much greater endurance. When a runner trains for long distance races, the long fibers develop preferentially. With a sprinter the converse occurs. To be best at a particular type of race, short or long, one needs the most of the short or long fibers, respectively, and it is impossible to maximize both at the same time. Other constraints exist as well. For example, the type of response that allows us to respond to emergencies (the fight or flight response) is damaging to the body if maintained constantly. Thus the type A person who seems to be getting so much done is not doing so without a cost.

We can see the role of "technology" in the way in which different metabolic systems or the construction of different biological structures have played such an important role in the evolution of life. For example, early ocean creatures were soft bodied or had external shells or bony structures. When bony fish evolved, they quickly became dominant. When woody plants evolved, they quickly replaced the earlier ferns and soft stemmed plants. New structures that evolve such as shells, bony skeletons, jaws, or woody stems may be compared to new technologies that humans invent, with similar consequences.

Information also plays a significant role. All life forms respond in some way to information by pursuing food, avoiding predators, seeking shelter, etc. But in particular, there are many cases where organisms use information to circumvent the limitations inherent in the physical structures of the organism. As a simple case, many plants respond to weather information that indicates the coming of freezing temperatures or drought by becoming temporarily inactive (dropping their leaves, etc.). Plants are typically also plastic in their growth responses, and will slow down their growth if resources are limited in order to persist with the resources available. This ability is what accounts for the existence of bonsai plants. If animals had this capability, then we would see elephants the size of dogs in adverse habitats, but we do not, though reptiles and fish are somewhat plastic in their growth. Plants can even adapt their morphology. In many species, the leaves produced in the shade on the lower branches are more adapted to shaded conditions (thin, unlobed) whereas those produced in full sun are shaped differently (thicker, deeply lobed). In this way information is used to adjust the morphology of the plant to enable it to survive a wider range of conditions.

Economic constraints and tradeoffs

In any economic system there are constraints and tradeoffs. Failure to remember this is a constant source of comfort to con artists and get rich quick schemers. One of the most fundamental is the tradeoff between risk and yield for investments. Stock market investors are often encouraged to take on more risk to get a higher yield. If the person plans to hold on to their investment for 20 years, this may be good advice, but otherwise "more risk" may mean in practice that their life savings could drop in value by 30%. When put in these terms many people are not willing to take any risk at all, and prefer an insured financial instrument such as a Treasury Bill or certificate of deposit.

A second economic constraint is that every action has a cost. This is an unhappy constraint for many people. The cost of meetings is rarely counted, for example, but can be quite high. New government rules and regulations (animal welfare, human subjects, accounting, safety, non-discrimination) have caused, over the last few decades, a doubling of the percentage of university staff devoted to generating the paperwork required to document compliance. This has increased overhead costs on grants which has caused other government agencies (those providing research grants) to complain about excessive overhead. The same thing has occurred in the case of unfunded federal mandates (rules from the national government that cause expenses at the state or local level, with no funds provided to comply with these rules), such as many environmental regulations. In each case, no accounting has been made of the costs of the requirements and whether the benefit justifies the cost. If personal life followed the same stringent requirements as dictated by the U.S. Occupational Safety and Health Administration (OSHA) and the Environmental Protection Agency (EPA), one would need to don a space suit to use ammonia floor cleaner or paint a bedroom, and all sports would be banned as too dangerous. Gilbert (1978) brilliantly characterizes the nature of the problem: the focus has been on regulating behavior and documenting behavior rather than on regulating outcomes. Regulating behavior always has a very high cost because one must be able to document compliance at the level of hundreds of individual actions. Regulating outcomes does not necessarily have a high cost and only aggregate statistics (e.g. injury rates) need be compiled. For example, OSHA regulates safety railing height, ladder construction, training requirements for ladder use, and a thousand other minutiae, and compliance must be documented, even for factories that are inherently safe or that have a well-trained and careful workforce. The outcome (lack of injuries) is what is important, not the record keeping. From 1988 to 1994, new U.S. Department of Energy (DOE) orders for safety accountability led to more than a doubling of safety department budgets at the DOE national laboratories (not counting huge expenditures for increased personnel training and safety meetings) with no improvement in actual safety (accidents and radiation exposure), according to the Galvin Commission Report to DOE (Galvin 1995).

The general tendency when faced with an economic constraint is to either spend someone else's money, or gloss over the hidden costs. In the examples above, regulators were spending other people's money. In a corporate reorganization or downsizing, the costs of disruption and lowered morale do not appear as columns in the accounting ledger and are thus often ignored. When shopping, most people do not view gas, auto mileage, and the cost of eating lunch out as subtracting from the savings resulting from going all over town to save $10 on a dress or bowling ball. Because it is a necessary constraint, it is useful to ask how much any action really costs and to be suspicious when someone acts as though an action is free.

Personal living constraints and tradeoffs

In personal living there are also constraints and tradeoffs that must be recognized. For example, because there are a limited number of hours in the day, people who try to "have it all" (a high powered job, a family, housework, hobbies, recreation, gourmet cooking, sports) find that they are exhausted all the time, and in fact are not giving fully in one or more areas (ignoring the kids or spouse, for example, or getting no exercise and therefore cheating their future health). Only the individual with extreme physical stamina can endure such a schedule. The same is true of the entrepreneur who puts in 70 hours per week. People make up all kinds of rationalizations to cover the gaps in their obligations, such as the myth of "quality time" and smoking to help them cope with nervousness.

In other areas of life also there are such tradeoffs. In romance one can have a number of casual dates or one can develop an intense relationship, but one can't do both at the same time. When driving, going too fast is dangerous, but the exceedingly cautious driver also is dangerous because they do not act decisively (hesitating in the middle of an intersection, for example). In the realm of careers, those who try to keep their options open too long end up without any marketable skill.

Unwillingness to recognize that such tradeoffs exist is a constant source of problems for people. A woman falls in love with an extroverted, athletic man and then resents that he loves to be away from the house playing sports and that he is not sensitive and responsive. A young couple will buy a huge house while both are working, and then when they start a family they will lament that they "have to" put the child in day care so the wife can work. When house hunting, people often want a combination of features that does not exist, such as a suburban style yard and neighborhood right next to downtown. Politicians hate to admit tradeoffs and typically end up "having it all" by borrowing and making the deficit larger. The simple reason for this is that for each member of Congress, the tradeoffs they see are different. One is willing to trade a smaller military for more education, another is willing to reduce entitlements in exchange for more medical research, but no one is in charge who can make them agree on what is being traded off for what; thus the default tradeoff is against the future by borrowing.

Although the opportunities are not as great as in technological innovation, in personal living also new technologies and information can help overcome constraints. For example, as much as day-timers and being penciled in on the calendar are laughed about, being organized (utilizing information) can help reduce the chaos inherent in a busy schedule. Information on nutrition and vitamins can help maintain stamina. Information technologies such as cell phones, answering machines, etc. can help in this area as well. New technologies help one to circumvent scheduling problems. For example, with a VCR one can tape a show when one is busy and then watch it when one is free. This overcomes the constraint that one can not be in more than one place at a time. Answering machines perform this same function (though they only annoy your friends if you don't reply to a message). Portable computers help make otherwise wasted time on a plane into useful time, thereby helping overcome the hours-in-a-day constraint. And so on. Thus we can see that strategic thinking applies directly in everyday life just as it does in one's profession.

Constraints: summary

There are always constraints and tradeoffs in any problem or system. Ignoring them is called wishful thinking. Strategic thinking requires a clear awareness of constraints. New processes, systems, materials, or information may allow constraints to be overcome, as in the gas mileage example. Awareness of constraints can tip one off to efforts to deceive, as when one is promised something for nothing. For some problems one actually works directly with the constraints to find a solution. In mathematical optimization, for example, one uses the constraints to bound the problem and to find a solution. When searching for an oil pipeline leak, one can be certain that the oil entering must eventually exit somewhere and thus flow rate changes can be used to localize the leak. Overall, an understanding of constraints is critical to strategic thinking.

Complexity

In real world problems, we are usually faced with extreme complexity. Whether it is predicting the weather or understanding the causes of urban decay, whether tracing the causes of divorce or puzzling over the spread of AIDS, we are faced with multiple pathways of causation, multiple actors, and various kinds of heterogeneity. The common response to complexity is to focus in on a single cause. Thus some insist that homelessness is strictly a function of poverty and resist any evidence that a large number of the homeless are alcoholics who will not take a job if one is offered. Divorce is sometimes blamed entirely on the feminist movement, when in reality several other social factors contribute to it. Such single cause analysis makes it easy to write letters to the editor of the newspaper, but is not really useful for solving the problems at hand. For this we must be able to disentangle complexity and be able to analyze it. This section addresses this issue.

Detecting patterns

A clue to the nature of complex behavior is the creation of patterns. These patterns can provide clues to the nature of the web of causation involved. Patterns can occur in space, in time, or in categories. For example, decades before it was understood that yellow fever and other diseases are transmitted by mosquitoes, it was noted that these diseases were most prominent during the summer. It was also noted that those living near swamps were most affected. The existence of a pattern in space and one in time helped point the way to the solution of the mystery of the source of these diseases which until this century had not only killed millions but had made certain regions barely habitable.

The existence of patterns was crucial to untangling the causation of AIDS. This disease was particularly difficult to comprehend because of its unusual difficulty of transmission and because of its long latency period. Disease experts usually must only trace back a few months to a year at most (in the case of the usual sexually transmitted diseases) to discover how contagion has spread, but AIDS takes many years to manifest itself. The fact that the initial occurrence was within a restricted population of gay men in San Francisco was what made it possible to track down its origin and method of spread.

The existence of spatial patterns can also lead to the uncovering of scientific principles. For example, the regular hexagonal pattern of cracks in dried mud flats indicates that there is a least energy principle at work, as does the regular configuration of soap bubbles on wire frames. The pattern of ripples in the sand below the shallow water at the beach also appears to be the result of some simple regular process, but surprisingly the mechanism has not yet been worked out. In biology, it has been noted that certain types of plants are only found in certain habitats (e.g. cacti in deserts, mosses in moist areas). These patterns have led to the discovery of how different habits of growth and physiological responses allow plants to live in different habitats (i.e. the patterns hinted at relationships between form and function).

Tracing the threads of causation

In many cases, causation does not lead to simple patterns in time or space or by discrete groupings. Rather, one must trace out the threads of causation, like tracking the course of a root through the soil. This is particularly so for distributed causation. A distributed effect is one where the causal agent acts at many points or through many individuals. In such a case there is no single place where one can stand and observe an effect. For example, we can stand and observe a person hit a baseball or directly observe hurricane damage, but the overfishing of a lake happens one fisherman at a time, gradually, so that it can not be directly observed. The difficulty with such processes is that we tend to ignore such little, incremental, or distributed effects and they thus become invisible. There are some general principles, however, that can be applied to help trace such effects and causes.

NOTHING IS FREE

Whereas it is obvious that nothing is free at the store, in other realms we treat certain goods as if they were free because the costs do not have to be paid directly. The realization that nothing is free provides a tool for tracing out consequences. Within organizations, there is a tendency to view meetings and requests for information as free. Clearly managers need information in order to manage, but the cost of obtaining this information is rarely accounted for. Even more costly is the time spent in meetings, not only in direct cost of time spent by all parties (up to $30–$100 per hour per person for senior people), but in space required for meeting rooms, in the costs of setting up a meeting, and in the costs of disrupted schedules, for a total of up to $2000 for a one hour staff meeting for a 20 person department. If meetings had to be purchased and approved the way equipment is, it is doubtful that they would be taken so lightly or that so many of them would be a waste of time.

A corollary of this principle is that virtue is also not free. Let us take as an example the Americans With Disabilities Act. This is certainly a virtuous law. At the level of requiring handicap access, we can clearly afford to have a few handicap spots reserved at the supermarket and to have sidewalks and doors that allow wheelchair access. These costs are negligible. But when carried to extremes, the law allows absurdities that interfere with the practice of every profession. A woman sued a movie theater because she was too fat to fit in the regular seats and wanted to bring her own chair, but her chair blocked the aisles so the theater would not let her in. A blind student sued a medical school to allow her in as a student, but they argued that it is impossible to examine patients without sight and certainly impossible to do surgical procedures without sight. When such suits become common (as they are beginning to) then they become a hidden, distributed cost of a virtue law that is too vague.

A major area where goods are assumed to be free is in the realm of pollution control. The pollution control laws as now written do not take

much, if any, account of the cost of reducing various pollutants or
cleaning up waste sites. If an acre of land costs $10 000, does it make
sense to mandate spending $1 000 000 cleaning it up if there is no risk
that the contamination will spread? There is an absolutist trend among
some in the environmental movement who insist on reducing pesticides in food and pollutants in the environment to zero, even though
the cost of doing so is infinite because it is impossible to perform any
action without creating some type of waste. The literal interpretation of
the Delaney Clause (no harmful pesticide residues allowed in food taken
to mean zero rather than very low and safe levels) becomes ludicrous
when we can measure chemical residues to parts per trillion, and will
cost the economy untold amounts in lost agricultural production.

In general, it is easy for those in authority to request or order
actions that are expensive if the costs are distributed and must be borne
by someone else. Congress is the ultimate example of this tendency.
This has led to the reaction by local governments against unfunded
mandates. Reorganizations, for another example, are also not free but
are often treated as if they were, and in fact can be disastrously
expensive.

EVERYTHING GOES SOMEPLACE

This seemingly obvious principle has significant consequences for
tracing out causation in various realms. Let us examine pets. All pets
in a pet store that are sold go someplace, usually a home. What then?
Many pets are kept until they die (which is sometimes a brief time!), but
what about the rest? Some escape and some are let go. For these, what
then? Where do they go? If it is a pet bunny, it probably gets eaten but
may join the other wild rabbits. However, if it is a walking catfish that
escapes, it may become the founder of a wild population. The consequence of ignoring the principle that everything goes someplace is that
dozens of species from other parts of the world now make their home
in Florida, including monkeys, poison toads, tilapia, pythons, and catfish. Wildlife officials are now much more cautious about imports that
have the potential to escape, though perhaps still not cautious enough.

Let us take as another example tires. All tires wear down eventually. On heavily traveled roads one can calculate that the amount of
tire dust is enough that the sides of roads should be inches deep in the
stuff. Since they aren't, we can ask where did it go? A little investigation
reveals that bacteria break tire dust down gradually. Without asking
where it went, we would never know this fact because the phenomenon is not visible.

This principle is most particularly relevant when tracing out the fate of pollutants. It is easy to assume, as was the wont in the old days, that once pollution blows away, it is gone. For bad smells, dust, and pollen, this is certainly true, because dust settles back as dirt and pollen decays after it hits the ground. For other pollutants, however, we must remember that it all has to go somewhere. For pollutants such as lead that do not decay, this means that eventually they will become a problem wherever they go, as has turned out to be the case.

ENTROPY IS OUT TO GET YOU

We may summarize this principle as that you can not unscramble an egg. In the realm of pollution control this means that it is impossibly expensive to go out and collect the pollution once it has been spread all over the landscape. In the realm of governance, this means that it is much harder to get good morale back than it is to lose it, and much easier to create anarchy than to undo anarchy (witness Bosnia, Haiti, and Somalia). Overall, entropy is the tendency for disorder to enter any system or structure. Metal rusts, books get dirty and faded, paint peels.

An understanding of entropy is very useful for tracing out causation. Entropy indicates that little glitches and imperfections are likely to accrue every time there is a transaction or a communication and also that there will be a cost to maintaining anything. What this means is that many types of problems result purely from the structure of a system. For example, negotiations between multiple parties are far more likely to be problematic when each speaks a different language and comes from a different culture purely because of translation and interpretation errors. One need not necessarily look for a "reason" that problems occur in this setting or ask who was the bad guy who scuttled the dialogue because noise alone may be to blame.

We may ask how some laws come to be interpreted in such strange ways and find again that entropy has a lot to do with it. Consider that in a legal system that gives strong weight to precedent, noise (e.g. a particularly persuasive lawyer, a judge with strong personal opinions) can gradually distort the intent of the law. Once a ruling is made in a certain direction, there is a tendency to cite this as precedent. Since no one tallies up all the rulings and summarizes the predominant interpretation of the law, unusual rulings can carry disproportionate weight. Thus these distortions (noise) can build up and either change how the law is applied or create hopeless confusion and uneven application of the law. Such is the case in product liability trials, in the treatment of drug violations, and in cases involving the criminally insane: very inconsistent, almost random application of the law. The fault lies in a system that amplifies noise by relying too heavily on precedent.

As another example, the more arcane a system, the more sensitive it is, and the more steps are involved, the more likely it is that entropy (noise) will be a problem. This follows directly from the multi-step problem solving model in Section 2.3 "Strategic problem solving." When I worked for an organization with a loose procurement system, entropy was not a problem. If there was a mistake on a procurement form, the procurement officer would fix it or maybe give me a call for clarification. Later, under a new set of strict rules, even trivial mistakes would send the whole thing back to be redone. When this occurred, one had to start over with the entire chain of signatures. Since the system had simultaneously become more complex (new rules, more rules, new forms), which caused more errors, and more sensitive (less tolerant of minor mistakes), noise in the system was not smoothed out but was amplified. The consequence was that the wait for purchasing a computer or even software became extended to 12 months, and service contracts could take 18 months.

Thus we can see from these examples that entropy is not merely something that happens to ancient pottery, but inserts itself into every process. If a system becomes hypersensitive to noise, then it can over-react or become swamped with delays (as above). The Swedish navy recently sheepishly admitted that their accusations of near shore approaches by Soviet subs were really misinterpretations of the sounds of swimming minks. Their acoustic system was hypersensitive, but had inadequate protection against false signals. Entropy can also lead to a gradual divergence between a system and its intent, as in the legal system example. In these and other cases, one must understand noise to understand how the system works.

Degrees of complexity

A very important type of pattern results from different types and degrees of complexity. If only a few objects or forces are involved in an interaction we can often predict the result. In the simple system we can trace out the effects at the level of the details (e.g. the paths of two billiard balls). Similarly, if very large numbers of objects are involved, the behavior may again appear simple or regular (e.g. the rate of popping of movie theater popcorn). A very large complex system can become simple because one observes the average or ensemble behavior, and not the details. Thus in a gas we do not try to trace out all the collisions or try to measure the speed of every molecule, but rather take the temperature, which is a measure of the average energy across all the molecules.

A totally different effect occurs when complexity is at an intermediate level. Such middle number systems are likely to produce complex and unexpected behaviors. For example, natural ecosystems are very complex and behave like large number systems, producing fairly predictable behaviors at an aggregate level (average plant mass, average number of species). If, however, we are interested in the dynamics of an individual species within this system, it may be subject to many forces and limitations such that it becomes almost impossible to predict the population changes over time. In very simplified systems such as in the laboratory, one again achieves predictability.

We may extend this analysis to national economies. Modern industrial economies are so complex (so many different industries, sources of capital, outlets for trade, resource bases, interactions) that no single economic problem affects the whole economy. In contrast, a nation that is just beginning to industrialize is more like a middle number system: complex, but not extremely complex. The result is much more unpredictability, more complex dynamics, and more volatility. This leads to more extreme economic crashes such as the Great Depression of the 1930s and the depressions of the late 1800s in Europe and America.

Thus it can be very helpful to know the type of complexity one is dealing with. For simple systems, observations or simple experiments may provide insight. Control is easy. For large number systems, ensemble behaviors are more useful measures and again control may be possible. For middle number systems we should be aware that predictability is more problematic and control nearly impossible. One should be particularly concerned if the system one is trying to create turns into a middle number system. Sophisticated systems analysis (simulation) may be needed in these cases. With such an analysis, one should seek to uncover modes of response (cycles, instabilities, chaotic dynamics or attractors, thresholds of response, buffering, etc.) rather than expecting to make precise predictions. In this way, the characteristic modes of behavior of the El Niño weather system have become understood, even though we can not make precise predictions about it.

Summary

When attempting to understand or manipulate complex systems, it is imperative not to allow the desire for simplicity to rule. Single factor explanations or policies are simply inadequate for such situations. Patterns can provide clues to underlying processes. Tracing out causation and deciphering behavior can be usefully achieved by understanding conservation laws (nothing is free, everything goes someplace) and the effects of entropy. An understanding of complexity per se is also essential.

Feedback and information

An understanding of the roles of feedback and information in complex problem domains is essential for effective problem solving. Feedback is a critical determinant of the behavior of complex systems. The way in which information is transmitted in a system must be understood because of the ways in which information is biased, blocked, and distorted. Finally, proper utilization of information feedback is crucial to enhancing personal performance as a professional. These issues are explored here in turn.

Feedback and system behavior

For any social, biological, or mechanical system, we must consider feedback to grasp how it functions. Without feedback to regulate such a system, it will fall into disorder and lose functionality. A person gets hungry and eats. If the feedback to hunger is destroyed (e.g. by a brain injury), the person will starve and die. This is a symptom of anorexia. During driving, one must constantly correct the course of the car to stay on the road, using feedback from the location of the car. In many types of systems, the feedback is not conscious as it is during driving, but results from the interplay of forces. For example, wolves are generally unable to reduce deer to too low a level because they become too hard to find. If deer get too common, they become malnourished, which makes them easier to catch, which causes the wolf population to increase, which reduces the deer population. This is a classic population interaction in ecology. The feedback here is not designed and therefore does not achieve a perfect balance. It is loosely governed between too few deer and too many and tends to cycle over this range. If wolves are removed (and human hunters are also removed), then deer increase to the point where they damage the vegetation and are at risk of massive starvation. Again there is feedback limiting the deer, but they are limited by starvation and not by predators.

In the wolf–deer system we can recognize several typical components of feedback systems. Positive feedback is the tendency of deer to increase more rapidly when they are low in number (because food is more abundant). This tends to help them avoid extinction. Negative feedback by either predation or starvation limits the maximum population size. A lag may be recognized in the time it takes the wolf population to increase when the deer become very abundant. If positive feedback is too strong or lags are too great, then there will be a tendency to overshoot and collapse. This can be observed in many insect populations whose natural rate of increase is so great (with each adult female laying hundreds to thousands of eggs, they can increase very fast). In such populations, boom and bust cycles are observed (e.g. locusts, gypsy moths, spruce budworm).

The examination of another natural system, the ant colony, provides an example of how even a collection of seemingly stupid actors can achieve results when guided by feedback. Ants are not too bright, don't see too well, and wander around a lot, yet they are very successful in every imaginable habitat except the Arctic. How is this possible? The answer lies in a series of pheromones the ants give off that coordinate their behavior. When an ant is out foraging, it wanders around quite a bit. If it finds a good source of food (a fruit or dropped candy bar) it heads back to the nest and lays down a special trail that alerts other ants that good food is out there. It does not make a beeline back to the nest, however, but follows its original trail back. This makes a quite crooked path for other ants to follow, but it may be observed that over time the path gets straightened as other ants come and go to the food. The reason is that the new ants gradually cut off parts of the trail that wander, and those that make a more direct path leave a stronger scent trail (because it is fresher); thus more ants are recruited to the straighter path. When the food is used up, the ants don't leave a scent trail when they return, and activity diminishes as the scent evaporates. Similarly, when an ant is attacked or killed, it gives off a special scent that attracts help. Each ant is a free agent, but is able to recruit others to help dig, fight, or forage if it encounters special conditions. Similarly, it may be recruited by others. These feedbacks governed by pheromones create concerted, though rather chaotic, action to serve the goals of the colony.

These same elements of feedback operate in any human system. They lie behind the laws of supply and demand, for example. If an item becomes scarce or demand increases, this calls forth higher prices, which stimulate more production of that good or service. This is a classic negative feedback. A lag can introduce economic dislocations. For example, if demand increases for copper, this stimulates mining efforts. However, mine expansion is typically a prolonged process, usually requiring several years. This introduces a lag, such that production of metals like copper tends to be out of phase with demand, causing strongly cyclic prices and periodic large layoffs among miners.

We may take as another example the governance of factory production rates. In some settings, inventory has governed the production rate on the factory floor. The production line would go full steam until inventories built up to an intolerable level, at which point production would be cut back. Thus using inventories to regulate production causes a lag to be introduced between when there is a change in the sales rate and when this becomes noticeable as a change in inventory. The consequence can be a cranking up (paying overtime) and down (layoffs) of the assembly line. Most modern manufacturing does not base production rates solely on inventory, for this very reason, but instead uses information on sales rates and historical data on how sales change with season (e.g. the intentional buildup of inventory for the Christmas rush).

When we attempt to solve problems or understand social systems, discerning the role of feedback and lags is crucial. A run on a bank and a stock market freefall are both positive feedback responses. Once a panic sets in, it breeds further panic, causing a run on a bank. The response during the 1930s of declaring a bank holiday was the proper one because it broke the feedback cycle by allowing people time to cool off. The same has now been introduced on the New York Stock Exchange by automatically shutting down the exchange for a period of time if prices fall too much too fast, which again allows a cooling off and breaks the positive feedback cycle. Positive feedback can also be implicated in the acceleration of hostilities leading up to some wars and the tendency of the media to focus on the same few issues at any one time; once a few cover it, others don't want to be left out, which makes it a bigger story.

The invisible hand of economics is basically the operation of negative feedback and positive feedback working in conjuction. Supply and demand are kept in balance by these opposing forces (at least in an ideal market). When these forces are prevented from operating, then distortions become evident. Thus in communist Russia in the past, lines at stores were long for many types of goods because prices and production were fixed and did not reflect demand. When a good is subsidized, it will appear cheap and may be overexploited, such as Western grazing lands at the turn of the century. In the case of the public grazing lands, there was no feedback about the increasing scarcity of grass (and thus of overgrazing) because it was not priced according to scarcity but was virtually given away. This caused large-scale range degradation and conversion to shrubs or weeds. In the Soviet Union, bread was subsidized to such an extent that it was actually fed to pigs instead of grain.

Lags in social systems can also cause odd behaviors. One example concerns the movement of people in the United States to different cities or regions. For many years the word was out that Los Angeles was the land of opportunity. Millions moved there. However, when the LA highways became gridlocked and the air became dirty and the jobs dried up, the word did not immediately get out. For at least a decade people kept moving there. It was not until the early 1990s when large numbers of Angelenos began to move away looking for work that the word began to spread that maybe it wasn't the best place to go. The lag involved in negative perceptions can be even more persistent. Cities that acquire a reputation as ugly or terrible to live in become the widespread butt of jokes, which continue long after the city has improved (Cleveland being a case in point).

Positive feedback can lead to seemingly inexplicable results in social systems. We may take as an example the economic phenomenon of increasing returns proposed by Brian Arthur (Waldrop, 1992). It has been observed that the product that comes to dominate a market is not necessarily superior in either price or functionality. This is quite contrary to standard economic theory which holds that consumers should make rational choices between competing products. Increasing returns explains this phenomenon. If a product is judged not just on its attributes, but on how abundant it is, then a product with an initial market share advantage may come to dominate even if it is inferior. When vacationers shop, they may prefer to go to the mall even though the stores are the same as back home, precisely because when the stores are the same they can return the items later if there is a problem or the clothes don't fit. Thus more common stores have an advantage.

We may take the personal computer as another example. When IBM introduced its personal computer, it was inferior to the Apple Macintosh in many ways. However, because IBM could afford to advertise on a large scale, they captured a significant share of the market. Once this occurred, software companies flocked to the IBM because it represented a larger market. Any new purchaser of a PC could then easily see that more software was available for the IBM-type machine than for the Apple. This plus issues of compatibility gave the more abundant machine the advantage. The same scenario has been played out for the 8-track versus VHS videotape technologies. Such dominance by an inferior technology is not necessarily permanent; if a new technology is vastly superior to an old common one it will still catch on. Increasing returns can also be observed in politics, where people are reluctant to join a small splinter party, and in ecology where a species that is rare may have an excessive disadvantage due to difficulty in forming flocks or in finding mates. Thus positive feedback can explain outcomes that otherwise seem to be exceptions.

In general, then, when solving a problem involving a social, biological, or engineered system, it is advisable to look for the feedback loops and lags that govern the system. These provide the greatest potential for understanding and control. Thus in the case of the stock market, a pause in trading (a lag) has proven sufficient to break the positive feedback of freefall. In the case of ghetto poverty, there is a strong positive feedback loop of widespread poverty causing conditions for children that discourage their success later. Attempts to alleviate this suffering in the short term (welfare) do nothing to break this strong positive feedback cycle. Likewise, when low income housing is built as a helpful measure, it actually amplifies the positive feedback cycle if the housing is large scale because this creates a large poor neighborhood which isolates children in this setting from the larger society. In arid grasslands, once overgrazing passes a certain point, the conversion to shrub lands (with low grazing value) is self-accelerating because of strong positive feedbacks in the system. Understanding this is the key to preventing such conversion because reclamation is prohibitively expensive. When managing any such social or biological system, the key is to use the strong feedback loops in the system to one's advantage, rather than to try to fight them head on. Thus in managing deer it is not useful to try to fight the positive feedback of population growth, but rather to manage hunting pressure, the negative feedback component. Those who oppose all hunting have no grasp of the fact that deer can strip every bit of vegetation from an area and then starve to death if there is no negative feedback (since wolves and cougars are missing from the system). Lags may be particularly good control points because they can often be overcome with information, as in the factory inventory example above.

It is also good to be aware that there are multiple feedback loops in many systems such that simple control points may not exist. In the deer example, there are several types of negative feedback that limit the population as it increases: increased predation, disease due to poor nutrition, disease due to crowding (easier transmission), and lack of shelter. In the economy there are also multiple pathways of feedback which can make the success or failure of a company far from predictable. This is also why we get seemingly inexplicable differences between countries such as the difference in extent of computerization of businesses between the United States and Europe. In such complex systems of multiple feedback loops, it is necessary to think a little more carefully and probably to diagram the causal feedbacks to get a grasp of the problem (see "Nets and webs" in Section 3.2).

Information transmission and distortion

Above, feedback was discussed largely in terms of processes, but information was also mentioned as a type of feedback. In human systems information is usually a key component of feedback systems, and thus the nature of information and its transmission needs to be clarified next. In addition, many problems result largely from failures of communication, and thus information also needs to be considered from this angle.

Information transmission is a far from perfect process with many opportunities for communication foul-ups. The basic steps are concept, encoding, transmission, and decoding. The concept step is not usually included but I wish to distinguish it here to point out that the original concept someone has may be quite fuzzy. It is not safe to assume that just because a person speaks that they know what they mean. This is the first danger in taking a legalistic or journalistic view of communication: that words actually represent the person's meaning. Many politicians have been pilloried for statements that may have really been half-cooked or even incoherent. Lawyers try to corner people with exact word phrasings, when actually they may not know precisely *what* they are thinking. The encoding step represents the conversion from a concept into some form of communication (speech, hand signals, flag waving, symbols). People have varying degrees of skill in this area. Most people have great difficulty expressing their thoughts, even if you give them time to sit and compose a letter. Extemporaneously, they may do very poorly. It is only the rare person who speaks in complete sentences and paragraphs. Often a verbal communication is quite garbled. Both presidents Bush had this problem. Such garbling of course becomes worse when one is under pressure (such as when answering press questions on TV) and distracted. This represents the second danger in interpreting a communication: assuming that the person was successful in expressing what they meant (if they even knew what they meant). This danger is constantly present when people talk about emotions (as in spousal "discussions") because (1) the person is likely confused in the first place, and (2) our vocabulary for expressing emotional states is highly inadequate. When a legalistic approach is taken to discussions of emotion, and a person is strictly held to something they said, then problems are only going to escalate. In such situations it is only safe to assume that no one knows what they feel and that they don't know how to express it. These problems are particularly acute in politics where people are held to the literal meaning of sound bites out of context. In legal settings the legalistic approach assumes that a witness knows what they mean to say, and say what they mean, with the consequence that if a witness can be tricked into a certain phrasing they may be held to it.

The third stage of communication is transmission of a message. In one-on-one conversation, transmission is not an issue. But for public communication it becomes a serious issue. The greatest difficulty is usually that complex issues must be summarized as bullets, sound bites, or 30-second summaries. It is very difficult to summarize complex issues such as international trade or a treaty on space exploration in a short bullet, and this leads to a serious loss of information that creates difficulties for public discourse. The second difficulty is that issues are often filtered through journalists, who introduce a bias, a point of view, that distorts the message in subtle or not so subtle ways. Finally, a message as it is transmitted may be inherently ambiguous. This is particularly so for messages encoded as symbols (a flag, a graphic sign, etc.) because such symbols, though compact, carry multiple meanings or only very vague meanings.

The final stage, decoding, is also a source of hidden trouble, and causes many problems. People usually assume that they understand correctly something they hear (i.e. that their interpretation is the correct one). There are many causes of decoding failures. A common problem is that the person does not have background knowledge in the area under discussion. A sports nut talking to his wife may encounter this problem. Cultural differences are a common cause of different interpretation of what one hears. If an American hears that a 30-year-old man lives with his parents, they will interpret this as immature behavior, whereas in many other cultures this is the definition of a good son. If the parents are ill, this may even be a *very* good son. There are also differences in perspective, discussed in Section 3.1, that cause a message to be given a different slant (e.g. the difference in how an optimist and a pessimist interpret what they hear). As a protection against such decoding failures, one should both be aware of the risk that one is misinterpreting a message and also be aware as a speaker of the differences in viewpoint on the part of one's listeners that may affect how they interpret what you say. In fact, it can be startling how badly people can misinterpret what they hear or read, even in the case of a scientist reading the work of colleagues.

Information feedback may also result from the compilation and analysis of measures of events in the world (consumer price index [CPI], sales figures, interest rates). Such information is abstracted from data which itself may be measured from various signs and symbols. For example, there is no one place, no dipstick, to obtain a measure of inflation which may be used as feedback for designing economic policies. Rather, inflation is measured as a composite of the changes in prices of many items. This example illustrates several problems with abstract information. First, just because you measure something and call it by a name, that does not mean that you are actually measuring something real or what you think you are measuring. In the case of consumer prices, many aspects of what is measured are misleading. If tires double in price over a period of time, we say that they have contributed to inflation. If, however, new tires now last 50 000 miles compared with 25 000 miles for tires from the earlier period, then they have actually not changed in price at all when measured as dollars per mile of wear. If, further, the new tires give better gas mileage and do not cause accidents by blowing out and causing a loss of control on the highway, then one actually gets more tire for less money than before. The same may be said for personal computers, new cars, VCRs, double pane windows, longer life shingles, longer life stain resistant carpet, and so on. Some new cars now can go 100 000 miles before the first tune-up. Airbags, which represent a real increase in value, are now standard on many new cars. There is a particular problem with comparing average home prices when people want bigger homes today and with car prices because modern cars have more functionality and get better gas mileage. Thus there may be serious biases in the CPI that cause it to be misleading, particularly when used as feedback to set interest rates. As an example, if people feel prosperous and buy bigger homes, this pushes up the CPI and leads to higher interest rates to "control" inflation which may not exist.

Second, the idea that one is actually measuring something real with an index is not necessarily supported by closer examination. For example, the concept that there is "an economy" is very misleading. Over the years, there has been ridiculous inflation of housing prices in much of California, particularly along the coast. For those who bought a house early and stayed, there was no problem. For those who hoped to move in from other parts of the country it soon became an insurmountable barrier. These two groups experienced entirely different "inflation." As another example, once one has bought a house and is getting older, the main components of inflation that are noticeable are college costs for the kids and medical costs. During the U.S. recession of 1992–4, it was even found that different regions of the country experienced recession at different times and to different degrees; thus in what sense is there a national economy?

When using information as feedback to guide institutions and organizations, one must take care that the measures are real and not biased or late. For example, one can have high customer satisfaction but still lose market share if no younger consumers pick up the item (i.e. all business is repeats). As another example, the problem was pointed out previously that if what is rewarded is working hard, then people will appear to work hard even if they produce little or nothing of value. This fallacy of rewarding behaviors rather than results (Gilbert, 1978) leads to inefficiency. It has recently been pointed out that many of the statistics that the government uses to guide policy are either biased (such as the CPI) or inadequate (Richman, 1993), which has significant ramifications for how government allocates resources and manages the economy.

Another example of how information is not derived in a straightforward way from data and may not mean what we think it does is the poverty rate. We often hear statistics about the poverty rate (x% of children live in poverty) that are used to drive government policy. The figures used to derive this statistic are rarely discussed but include certain simplifications that make them less than obvious in their meaning. For example, the fact that almost everyone starts out with a low salary and works their way up is not factored into these figures. Let us see what the consequence of this is. Consider a population that finishes high school at 17, works until they are 67 without ever being unemployed, and is guaranteed a 5% raise every year (without inflation). They begin work at the minimum wage of about $10 000/yr, which we all agree is not a living wage, and work up to $71 067, with an average national income of $30 533. Under this not totally unrealistic scenario, it takes nine years of annual raises before the entry level worker exceeds the poverty level of $15 000/yr. Under these conditions of guaranteed work and guaranteed raises, we still have 18% of the population living under the poverty line. At the same time, these young workers who are just starting out do not consider themselves poor (consider most college students, for example). They do not need much to live on because they are not yet trying to buy a house, do not mind having a room mate, are not yet trying to save for their kids' college fees, and may have been given a car and used furniture by their parents. Conversely, couples starting a family often find themselves in financial straits because they have medical bills and must get a larger house, bigger car, and new baby furniture all at once. Older people on a low income are not necessarily in financial distress because they have paid off their house, have investments, and are not trying to save for their kid's college or for retirement. Thus choosing a dollar cutoff line and declaring that those below it are in poverty is really meaningless and does not measure who is really in financial distress. The 40 year old who makes $28 000/yr is in much more serious shape than the 18 year old making $17 000/yr because he will be unable to support a family or save for retirement. Thus the use of the numbers below the poverty line as information feedback on the state of the economy to guide policy is virtually an exercise in throwing dice when such a simplistic measure is used.

In summary, information has a significant role to play in problem solving. In understanding a problem, it is not always best to assume that the problem is merely a function of concrete things or actions. It may well be that the way information is exchanged or the way data are converted into information actually is the cause of the problem. For example, much of the bad news about the environment, such as the rate of desertification in Africa, is reported uncritically by the media, when actually such news is often unsupported by the facts (Simon, 1980), and as such is a misleading guide to what actions governments should take (i.e. the "information" is actually part of the problem). What if the major portion of the "inflation" in recent decades resulted from trends toward buying bigger houses and nicer cars and fancier clothes? These signs of prosperity (or at least of choices) would then have caused efforts to choke down the economy to prevent "overheating." I am not arguing that this has occurred (though Richman [1993] does so argue), but that potential biases in relevant indices could indeed have led to this result. In this type of setting, what may be most needed is not better control or fancy policies, but better information. Options here include provision of more realistic conversion of data to information, faster feedback of information, reduction of noise in the signal, removal of blocks to information flow, providing assistance in information encoding or decoding, and disambiguating symbols. Loehle (2006) showed that applying these ideas to ocean fisheries by including expenditures on information as part of the economic and decision system can lead to greater stability of the fishery and higher overall yield.

Feedback and performance

If you were blindfolded and told to walk in a straight line, you would not be able to tell whether you were going straight or not. In fact very few people can go straight when blindfolded. Clearly we need feedback on our performance, but in the world of work, scientists and others may get very little feedback. Gilbert (1978) documents that for average performers in many settings (sales, factory work, etc.) the simple provision of information on what performance (in terms of products) is expected and of rapid feedback on how one is doing at any given time can substantially increase performance. Critical here is that feedback should include information on what specific aspects of output (defects, timeliness, numbers, quality, innovativeness) are deficient. An annual performance review is totally incapable of providing this type of feedback. With such feedback, one can continually learn and improve. Gilbert contrasts this approach with the more common one of rewarding behavior. Managers typically reward two types of behaviors: hard work and loyalty. But hard work, manifested as long hours, rushing around, staying at the desk, and taking work home, may have little relation to performance as measured by value added or return on investment, as Gilbert documents. Such excessive displays of hard work may actually be indicative of inefficiency, disorder, and problems with the system itself. A "hard worker" certainly has no time for creativity or learning. A hard worker, however, does demonstrate to the boss his eagerness to please and his subservience to company goals. Loyalty is manifested as enthusiastic support for the company, for the corporate culture, and for one's boss. Loyalty is valuable to a boss because it enhances his power base. Again, loyalty has nothing to do with performance and may even hurt it if loyalty prevents disagreeable things like innovation and needed changes in policy. The result of rewarding hard work and loyalty is that people find things to do to look busy and make a show of being loyal, because they respond to what they are rewarded for and what they obtain feedback about, but this type of feedback does not increase their performance or their value to the company.

We may compare several professions to observe the effects of performance feedback. When American auto makers in the 1970s began to be faced with Japanese cars with better gas mileage and higher quality, at first they fumbled and lost market share, but subsequently they recovered and have gone after quality. There are now a number of measures of performance by which they can demonstrate that their skill at making cars has gone up: gas mileage is up, exhaust emissions are down, consumer complaints and recalls are down, sales are up compared with Japanese cars, and the cars last long enough that some offer 60 000 mile or longer power train (parts) warranties.

Particularly interesting is a contrast between surgeons and general practitioners in the United States. A surgeon faces several types of feedback that keep his skills sharp and prompt continued improvement in technique. During surgery, he immediately sees if his diagnosis is wrong (the spot on the lung in the X-ray turns out to be nothing) and must with some chagrin sew the patient up without doing anything. People complain bitterly about being cut open for no reason. A mistake in surgery results in immediate problems such as spurting blood, loss of blood pressure, or damage to some other organ. In the case of a mistake, there are various people in the operating room who will notice and point it out. If the mistake is serious, the patient dies, which is pretty strong feedback. Following the surgery, the doctor visits the patient and obtains feedback about recovery, whether a fever has developed, whether symptoms have been alleviated, etc. It is often the case that an autopsy is done on any patient who dies during or following surgery, which provides further information (did they have a weak heart or did the doctor sew them up with a cut blood vessel?). If a doctor loses too many patients, the hospital may revoke his privileges. Thus a surgeon receives rapid, frequent, and emphatic feedback about the outcomes of his work, which enables him to improve his performance by finding out what works and what doesn't. Surgeons are sometimes held to be a little arrogant, but it is also true that a surgeon can obtain proof that he is in fact one hot shot performer. On the other hand, surgeons do not obtain very good feedback on the long-term effectiveness of their procedures. They generally work on the basis of their own experience which is largely anecdotal and does not control for the rate at which patients would get well on their own or for how long they would have lived without the surgery. Many surgical procedures have never been tested (performance feedback) in a controlled study, and when they are they often turn out to be ineffective, such as tonsilectomies, which are now rarely performed (Friend, 1995). Uncertainty remains about treatments such as stents for opening arteries in the heart and back surgery that can not be resolved by a surgeon's own experience.

We may contrast this situation with that of the general practitioner. The GP generally faces people with only moderately serious illness. If the person is really sick the GP will refer them to a specialist, but they need not really know what is wrong with the person they refer – a guess will do (I think you have an ulcer, go see a gastroenterologist). They will only rarely hear back about the true diagnosis of a person they refer to a specialist, so there is no feedback to improve their diagnostic skill (though under some health care systems they may receive such feedback). For less serious illness, they generally prescribe something and send the person away. For many types of illness (colds, flu), the person will get better even if the doctor provided no help, and thus will not return. Thus the doctor thinks his or her actions were helpful when the person doesn't return, when in fact the person may not return because they got better on their own. If the doctor was completely wrong, the person may look for another doctor or seek out a specialist on their own. The doctor usually does not enquire about the fate of patients who don't return (are they just healthy? moved away? died?) and thus never learns about patients who leave to find another doctor. There is no apparent effort to follow up on patients or to see how a sample of medicine performed. Overall, there is very little information at all fed back to a GP to help him or her hone their skills or learn about their deficiencies.

We may note another aspect of the feedback process with respect to performance in the domain of medicine. In a system with some measures of performance, the consumer can improve the system by making choices. In the case of automobiles, buyer's guides provide information on recall statistics, gas mileage, consumer satisfaction, resale value, and repair expenses for various models. After a while, a lemon will be either fixed or withdrawn from the market because sales fall. But for doctors, almost no information is available to the patient. With some difficulty you can find out where your doctor went to school and what his or her specialty is, but it is almost impossible to find out about death rates during surgery or even number of malpractice suits filed against the doctor, much less his or her accuracy in diagnosis. When malpractice suits are settled there is usually a clause of nondisclosure. If patients could find out more about their doctors, there would be an incentive for the doctor to improve his or her performance, and the overall system would improve as poorly performing doctors lost patients or quit practice altogether.

Let's look at some other professions. In sports, there is instant feedback every time an event is run or a game played. There is further feedback in the form of video playbacks. In this way the athlete can see exactly what he did wrong, with the help of his coach ("you held your head up there son, keep it down"). A consequence is that even the average pro athlete can achieve performances quite close to the best in the world. An active scientist receives feedback every time he or she writes an article or proposal, with detailed (even brutal) critiques from reviewers. These critiques often point out the specific things (lab technique, statistical analysis, etc.) that were done wrong, and thus help improve his or her skill. In teaching, in contrast, it is much easier to tell if you are keeping order and the student's interest than to tell if they are learning more than last year's class did. Gilbert (1978) discusses systems that could provide such feedback in education, but they are currently not much utilized.

Thus we can see that tremendous potential exists for improving performance in any field with feedback. In manufacturing tasks, output can be doubled or better when feedback is provided (Gilbert, 1978), but in more intellectual professions the potential is even greater. For example, the average academic may publish a few to 20 papers over a career, in contrast to the most prolific with over 1000 publications (see Section 2.3 "Strategic problem solving"), a ratio of 100 or more. The same applies in the area of inventing. Without feedback, it is very difficult to tell what actions lead to improvement and it is far too easy to drift. Feedback on work quality and status is particularly critical for highly iterative tasks, such as writing, software development, and design work. In all of these domains, one must almost always start with a rough product and successively refine it. Good feedback (testing, self-evaluation, peer review, etc.) can greatly enhance the iterative refinement process and make it more efficient.

Information and discovery

There are cases where the role of information can itself be an object of study that leads to new discoveries. This is particularly so in scientific research but is also true in business. If an actor takes some action that indicates awareness of information to which the actor does not have access, this indicates a mystery or a hoax. We can find examples from economics, extra sensory perception studies, animal behavior, and scientific fraud. Thus the tracing out of information flow can be useful for both discovering new facts and for disproving false ideas.

In economics, a long-standing simplification has been the omniscient optimizer as fundamental economic actor. This ideal actor would always buy the cheapest good or take the most efficient action. The body of economic theory built on this assumption was truly impressive, and was quite mathematical. However, a nagging suspicion developed over time that real people were not like this ideal economic agent. If you go to the department store and find the $4 steak knife you were looking for, you do not know for sure that this is the lowest price, but you are not likely to go all over town comparing prices to save $0.30 on this knife. Thus you must guess, based on general rules of thumb about prices of goods, whether this price is low enough. That is, you do not have perfect information. It is not the case that this is a trivial flaw in the ideal agent theories, because it leads to significant economic behaviors. For example, the success of such major chains as Target and WalMart results from the fact that on average their thousands of items are priced low enough that it is not worth comparison shopping: you can go in and buy dozens of items and be pretty sure it would not have been worth shopping at a bunch of specialty stores, since this would take a lot more time. The fact that the shopper is not an ideal agent also explains aspects of brand loyalty (too much trouble to check out all competing brands) and the occasional success of inferior products when advertising provides the dominant information about a product and it is not worth testing all the available products to find the best.

Animal behavior provides some interesting examples of information playing a central role in discovery. It was mentioned above that ants show organized behavior in spite of not being too bright. This fact led to the discovery of the first pheromones, because it seemed logically impossible that ants knew what they were doing and got together in little committees to plan their foraging expeditions. It was simple to show that crushing a few ants could start a panic (pheromones released), whereas the removal of some ants was blithely ignored (no pheromones). As another example, bird migration has long been a puzzle. How in the world can birds migrate such long distances without a map? Further, they even do it over open water and in the dark and with cloudy skies. Since there must be some information they utilize, many studies were done to uncover how they do it. One hypothesis, that birds can utilize the Earth's magnetic field, were at first met with incredulity, since of course magnetic fields can not be detected by animals. Studies have shown, however, that even bacteria have magnetite bodies in their cells that allow them to orient to the magnetic field of the Earth, and that animals have groups of cells in their brains that serve a similar purpose. Other studies appeared to show orientation by the light from stars which indicate true north or by other mechanisms. The contradictions between these results have been resolved by studies showing that birds probably use several mechanisms and intercalibrate them. Thus the migration of birds provided a puzzle that led to the discovery of new sensory mechanisms. It is worth noting that this is also an example of where either-or debate is not useful, because in this case the correct answer was that all of the proposed mechanisms are probably used, and not just one.

A similar problem has arisen in the study of altruism in animals. Social animals are known to exhibit behaviors that are inexplicable from a purely selfish point of view, such as standing guard or grooming each other. The theory of kin selection argues that if the others that an individual helps are related to it, then the individual can increase its fitness in an inclusive sense even if its action has a cost to itself. This theory works out genetically in a very convincing manner, and applies quite clearly in social groups such as a pride of lions. However, the behaviors in question also seem to occur in cases where the individuals do not have access to information about who is related to them. This is reminiscent of the omniscient economic agent discussed above. The solution (Pfennig and Sherman, 1995) appears that there are biochemical and other signals of relatedness, such as scents transmitted in urine, that animals can use to assess degree of relatedness. In other cases, such as many primates, individuals keep track of who has helped them in the past, even if these are not relatives. Once again, a paradox of information provided an open question that led to the discovery of a new mode of information transmission.

Extra sensory perception (ESP) is a case where information transmission still remains problematical. Here again, the putative phenomena involve transmission of information by an unknown vehicle. We might argue that if birds can detect magnetic fields to migrate in the dark, why could not humans transmit information across space to one another? The difference here, of course, is that no one can deny that birds migrate at night by quite straight paths, whereas concrete proof of ESP is not so universally accepted.

Summary

In summary, information and feedback are critical components of any dynamic system. Positive and negative feedback cycles actively govern dynamic systems and prevent collapse. Lags in feedback make achievement of optimal performance much more difficult, and generally result in oscillatory behavior. Information is a critical type of feedback, but unlike feedback in physical systems, information is not necessarily unambiguous. Information may be garbled, biased, abbreviated, or inadequate. Thus the most cost-effective strategy for improving system performance may simply be the improvement of information quality and flow, particularly for individual performance.

Bottlenecks

Every complex system has bottlenecks. Identifying them can be very useful for solving problems. An obvious bottleneck occurs in a highway system under repair. The bottleneck can be particularly acute when road system design is such that no alternate routes have been provided. Highway systems may even create their own bottlenecks by cutting off side streets, thereby making traffic more dependent on them. Transportation networks also are subject to paradoxes, such that attempts to fix traffic jams by adding capacity have no effect or actually make things worse (Arnott and Small, 1994). For example, various forms of mass transportation are generally more efficient and faster for each individual if more people take them (with more traffic, trains can run more often, reducing waiting time). The paradoxical effect of adding more road capacity can be that overall average commute times can go up as those using mass transit experience increased delays and switch to driving because there is insufficient ridership to allow trains or buses to run frequently (Arnott and Small, 1994). Road capacity increases can also paradoxically encourage people to take jobs farther from their homes and thereby become dependent on major arteries (highways) to get to work, where before they worked near home and took side streets.

It is a human tendency to hate limits and to try to push them back, but sometimes bottlenecks are useful. For example, in San Francisco traffic flow in and out of the city is very limited by the bridges, especially during rush hour. This means that it is difficult to live outside the city and commute. The consequence is that there is much less suburbification than in other cities. The central business district of San Francisco has thus not been abandoned. A city that is confined by such geographic circumstances is much more able to sustain mass transit, with San Francisco and Manhattan being good examples. Thus a traffic bottleneck can have a positive effect.

In a business context bottlenecks can be expensive and are often associated with control points. Any attempt to exercise oversight has the potential to create a bottleneck. If a manager must sign for every purchase then he must be very certain that he reviews them quickly to prevent a delay, because purchasing delays can cause production to stop or activities to be curtailed. If procurement decides to order items in bulk to save money and holds orders until enough accumulate, then one may have to wait six months to obtain a computer for a new employee, thereby causing a very expensive loss of productivity. Clearly this is foolish because if it is known that a certain number of computers are being ordered per month then a bulk order can be placed in advance, and waiting times can be reduced to weeks with the same cost.

Decision making often represents the most serious bottleneck. If too many signoffs are required for a new product, for example, then the delays introduced can be disastrous. This was the case with the old Xerox, which had so many signoffs required that some potentially major products never made it to market. We may contrast this with the culture at 3M, where an individual engineer can take initiative and develop a new product with no external delays. Decision points after a new product is developed are specific and involve limited numbers of decision makers. It is characteristic of 3M's product line that many of their individual products do not require huge startup capital (e.g. post-it notes, scotch tape), so a decision bottleneck is not useful. In contrast, a decision to develop a new model of automobile should not be taken lightly, and bottlenecks should actually be introduced if not present so that actions can be considered carefully.

It is characteristic of some bottlenecks that when their capacity is exceeded they create delays in a nonlinear fashion. We clearly observe this on a freeway, where traffic is either moving at the speed limit or crawling, but rarely can be observed moving smoothly at moderate speed. The same phenomenon can be observed when managers become overloaded, because then they feel overwhelmed and are reluctant to make major decisions. Under these conditions, additional time is spent shuffling papers and reprioritizing tasks, and so even less throughput is achieved than normal, causing even more work backup.

Bottlenecks need not be physical, but can be abstract. For example, in a study of a region of East Africa where there are both pastoralists and farmers, it was found that economic development was limited not by the lack of wealth but by the lack of money. That is, there was no reliable banking system. Without currency, it is not possible to save profits when one has an excess. For pastoralists, this means that wealth can only be saved in the form of live cattle, but live cattle must continually eat to be maintained. This means that the profitability of the system is much lower than if excess cattle could be sold and the money put in the bank. For the farmers, excess food that can not be eaten is simply waste. The lack of currency also does not allow anyone to borrow to try to improve efficiency or productivity. Thus currency in this system represents a bottleneck, but one that depends on social stability and institutional infrastructures.

For understanding a complex system, it is very useful to first identify the bottlenecks. These represent points where control may be exercised. It is not, however, sufficient to merely eliminate a bottleneck, because this is likely to merely shift the bottleneck somewhere else. Without knowledge of the entire system, the elimination of single bottlenecks can have unforeseen consequences and puts the manager in the position of the game at the carnival where one tries to whack the moles that pop out of the holes. Such a game has no resolution and commonly goes by the name of fire fighting in business. A properly designed system does not have any critical major bottlenecks that can back up production or work flow, but may have many little ones that cause only minor delays. A key to such design is to avoid flows of materials or decisions that are too centrally focused. Thus in the 3M example, there is no central place for new products, rather new products originate in a distributed fashion by the initiative of engineers. This ensures a continuous flow of new products, which has made 3M so successful. In distinct contrast, when a chief executive officer is a micro manager, he creates a severe bottleneck that can quickly stifle a corporation. On an assembly line, care must be taken that a single local problem with one car does not create a bottleneck. One notes that efficient assembly line procedures are set up such that one is never left waiting for parts. In other types of operations, it does not always occur to people that engineers without computers and workmen without tools (orders placed months ago) are "waiting for parts" just like on an assembly line, which has a cost. Thus every effort should be put into eliminating bottlenecks that involve parts, supplies, purchasing, etc. because these inexpensive items can cause large hidden expenses and such bottlenecks do not represent useful control points. A useful control point is one that provides oversight that prevents the hasty or uncontrolled procurement of buildings or startup of new factories. A distributed, flexible, matrix organization has the advantage that there are many small bottlenecks but few large ones. One does not obtain economies of scale in such an organization (which may be a myth in any case for modern manufacturing) but as compensation, delays and losses due to bottlenecks are reduced.

In design problems, bottlenecks are a constant source of problems and solutions. In computers, for example, bottlenecks in data flow can make or break a system design. A fast CPU is not much use with a slow bus, because the user will perceive it as a slow computer. As mentioned previously, highway design is often motivated by efforts to eliminate bottlenecks, though quite often the initiative comes from the public or politicians, and the counterintuitive aspect of network properties (increased throughput on highways creating the potential for more serious bottlenecks) may not be appreciated.

It is often the case that bottlenecks exist in information flow. Let us consider media reporting on complex issues such as the North American Free Trade Agreement or a tax bill. Such a bill or treaty is a complex document with many ramifications, but the only way that most people find out about it is from newspapers. This is a bottleneck of information flow. The reader rarely gets an objective brief summary of the document, much less the document itself, but rather is presented with reactions to it by commentators. This makes it very difficult to assess what the document actually says. A few years ago, the U.S. Federal Register was put on the Internet so that it may be accessed by anyone. This clearly removes an information bottleneck and has important consequences. It allows many people with an interest and expertise in the area touched on by federal actions to become informed before it is too late and is signed into law. Another area where information bottlenecks are important is in military operations. In the Vietnam War, the use of the jungle for concealment and tunnels for hiding troops prevented information on enemy activity from being obtained by the U.S. army. It thus provided an insurmountable information bottleneck that contributed to the withdrawal of American troops.

It is notable that the deliberate creation of information bottlenecks is a central strategy of most oppressive governments. Literature from outside the country is prohibited. Media are tightly controlled. Even fiction is restricted to certain themes. These actions show how an information bottleneck can be a control point.

Overall, bottlenecks are very important control points. Manipulating a system often involves the creation of control points in the form of bottlenecks. If bottlenecks already exist, manipulating them or relieving them can change system behavior. It is not the case, however, that it is always wise to eliminate bottlenecks, though the natural tendency is to do so.

4

The social dimension of science

Science is not merely conceptual, but instead involves active manipulation of the world to arrive at new knowledge. In this chapter, the implications of the ideas presented in the first chapters are explored. First, the process of developing and executing a scientific research program are explored, publishing papers is discussed, some ethical issues are touched on, and finally, the problem of interfacing science with the world of policy is addressed.

4.1 CONDUCTING A RESEARCH PROGRAM

Research that is random and aimless is unlikely to be very productive. In the age of naturalists, it was productive to collect specimens and observe animals in the wild because little or nothing was known. But today, any random observation is likely to be already known. Thus a key to success is to focus on a problem.

Earlier chapters discussed concepts that can help one to identify a problem worth working on. To review briefly: paradox can help one identify an opportunity; new tools and methods can open up topics to analysis; obtaining a larger data set or one at a larger scale than previous studies can lead to insights; and refining or testing a theory is always a productive avenue. Of course the idea of leverage as embodied in the Medawar zone can help one pick topics with a high payoff.

Given that a promising topic has been picked, the textbook view and the impression given to graduate students is that you should next design and conduct an experiment or study. However, successful scientists instead conduct research programs. The difference is that a research program is a sequence of studies designed to answer a question. This approach is needed because most topics must be dissected and the pieces of the puzzle solved one by one. For example, let's say one wishes to study a species of mouse in the field. In order to estimate their population one may encounter difficulties with catching them, which leads to a study of trapping methods and the statistics of sampling. Trying to determine what they eat may lead to a study of methods for analysis of stomach contents or the use of tracers such as stable isotopes. A controlled lab study may be needed which leads to all sorts of issues with lab artifacts, raising captive populations, and so on. At any point, the scientist may find himself needing to learn a new method from the literature or developing and publishing a new method if existing methods seem inadequate. Much of the literature in fact consists of development of methods, from statistical tests to lab procedures. The trick is to dissect the problem properly so that results of early studies lead to and facilitate later ones.

Types of studies

Science does not consist of a single activity or type of study. Fields differ widely in their tools and techniques as well as in how studies are carried out.

In physics a highly deductive method can often be applied. A theory is proposed which is very mathematical, such as Maxwell's equations or quantum mechanics. Sometimes existing data can be compared with this theory. If experiments are needed, there may be considerable technology involved, such as atom smashers or neutrino detectors. As well, software is often needed to filter and analyze the massive data sets produced. This level of experimental complexity can make progress slow and expensive. In astrophysics, say concerning the dynamics of the sun, experimentation is not possible and it may be difficult to tell what data would lend support to or refute a particular theory. String theory is particularly at risk due to a very tenuous link to experimentation.

At the opposite extreme, in psychology the theories that are proposed are quite vague and subjective. For example, hundreds of books offer advice on marriage or depression or whatever, based on the experiences of clinical psychologists. All of them have some sort of theory about the mind and emotions which they support with case studies, but the different books are often contradictory. How can this be? Several factors are in play. First, those who are motivated to come for therapy (and stick with it) are generally more motivated to change. Second, just having someone to talk to can give some people the courage to deal with their problems. Third, many times people come to therapy in a crisis but the cause of the crisis may become less intense over time, thus allowing them to get better independently of the therapy. The combination of these factors can give the impression to the therapist that he or she is being effective and that therefore the therapy is valid.

Psychology also provides examples of cases where new approaches and methods have allowed breakthroughs to be made. The Behaviorist School founded by B. F. Skinner famously declared that the emotions were off limits to study because they are subjective. This subjectivity seemed to be an insurmountable barrier until recently. It was noted that mirroring of the facial expressions of others seemed to be related to the proper interpretation of facial expressions and therefore of emotions. High speed photography has helped to document micro-movements (e.g. the start of a smile). Brain imaging has confirmed that motor neurons are activated when we see someone's face and the neurons activated correspond to the muscles one would use to mimic the face being observed. By imitating the face, even if tentatively, we can better feel what they feel. This was confirmed when it was shown that administration of a shot of a temporary paralysis drug around the mouth made it harder to interpret certain facial expressions because the subject was unable to mimic what they saw. This breakthrough was enabled by inventions and technologies from outside psychology per se. In addition, success was achieved, not by asking "What is emotion" but rather focusing on the communication of emotional states. In this field, it is very important to guard against subjectivity.

Archaeology is an interesting case where the objects of study are objective rather than subjective (e.g. pottery, buildings) but the interpretation of these objects has often been highly subjective, and therefore contentious. Consider the excavation of an ancient town. From the layout of the town and major buildings plus the tools and other items found there, a story can be built up about the people who lived there. This story can be pretty sketchy and it is thus easy to add interpretations that are not warranted: that life there was hard or easy; that the city was a regional center or not; that there was a strong priestly class or not; that women played a central role or not, etc. Without written records, other data are needed to squeeze some of the subjectivity out of the story. If ornamental objects are found that can only have originated in some distant place, then trade or conquest are indicated, but one can't tell if trade was by intermediaries, if the locals went there, or if the distant people came to that town. Thus jumping to conclusions can be dangerous, but is tempting.

It might seem that in anthropology it is not possible to do experiments, lacking a time machine. However, experimentation has played a big role. Ancient farming practices have been replicated, based on excavated tools and artwork, to see how time consuming it was and how much food could be raised. The manufacture of Paleolithic stone tools has been replicated. Ancient boats have been built, based on artifacts, drawings, and written descriptions and using only tools available during those periods, and tested for seaworthiness. With these ships one can evaluate how fast they could go, how many oarsmen were needed, and so on. Various ideas about how primitive people could have moved and raised stone monuments have been tested experimentally.

Whereas in physics a more hypothetico-deductive method can be applied and in anthropology we are trying to construct coherent stories about past societies, trade, migrations, etc., in some fields the question being asked is purely empirical. For example, in agriculture one may wish to find the most cost-effective level of fertilizer to apply or test whether a particular pesticide is effective against a particular pest. In engineering it may be that the breaking point of a particular structural element needs to be determined or rates of corrosion under particular conditions. In such applied sciences we are not necessarily either developing or testing a theory. Instead we are either conducting a pure statistical test, or developing a dose-response relationship (e.g. kill rate versus level of pesticide). Precision may be needed, which means good technique and large sample sizes are required, but perhaps not the level of rigor required for testing a law of nature.

Analysis of results

In a classroom lab exercise, the well-defined problem has a known answer and the experiment is set up to illustrate some point, such as yield from a chemical reaction or measurement of acceleration. In the real world, of course, scientists don't know the "correct" answer, and thus analysis and interpretation of the experiment is far more complicated.

Let us consider applied (empirical) studies first. Let us say that fertilizer is applied at several levels to several replicate fields and crop response is measured. If the experiment works out, the result is a regression relationship which allows one to interpolate between the tested levels. This relationship will have defined error bars. The optimal level of fertilizer to use will depend on the cost of fertilizer.

However, it may be that the variance between fields within each tested fertilizer level is such that the R^2 on the regression is very poor, which means one would not want to base management on such a relationship. How could this happen? It could be that the fields were not really comparable and differed in soil type or moisture status. This problem might be overcome by modeling ratio of fertilized yield to unfertilized yield (same field basis) to help control for nonuniform soil conditions, rather than modeling absolute yield. What if this doesn't help? There could be other uncontrolled differences between the fields such as disease. Or perhaps the fields have residual nutrients from prior years. There is even the possibility that your technicians (or you!) mixed up the samples! The failure to obtain a good relationship does not necessarily provide any clues as to what is wrong, as illustrated by this case.

In retrospective studies it can be difficult to obtain unambiguous results. For example, consider a study in which participants fill out questionnaires about their diet and health habits, and this is compared with their health profile. There is the obvious problem that people are loathe to admit their bad habits on a survey and will exaggerate how much good food they eat. Of *course* they all eat vegetables! This will make it harder to detect health effects. Let's assume that the surveys are done honestly and we find that those eating fish are healthier and even report less depression. Perhaps only Catholics can honestly say they eat fish every week (on Fridays) and it is known that religious people tend to suffer less from depression. If true, then the fish–depression link is entirely spurious. This illustrates a different type of confounding. Other examples of confounding include contamination of cultures in microbiology, inadvertent DNA transfer in a genetics study, and the fact that pure-bred lab rats are particularly prone to cancer.

How does one determine that a negative result (too noisy to be useful) is real? If the study was conducted by a student, it is reasonable to suspect sloppiness. Another possible culprit is too small a sample. This problem often arises because of limited time and money. Returning to the agricultural example, since it may seem like the level of fertilizer *must* be important, one could take the experiment to the greenhouse and use artificial soil and a randomized block design to eliminate sources of variation.

When experiments are intended to test a theory, the clearest test occurs when new phenomena are predicted. For example, the prediction that antimatter should exist is a very distinct prediction. Likewise, the effect of the Moon's gravity on Earth tides can be predicted and data confirm the prediction. Often, however, alternate theories or models make very similar predictions. For example, it has been observed that in any given locality there tend to be many rare species and a few abundant ones (this is called the species–abundance relationship). There are many models (curve shapes) that can mimic this type of distribution, and some of these can be argued to be justified by theory. The difficulty is that data to test them tend to be noisy and sample sizes inadequate to get a precise fit to any of the models. If you sample the same place over time to get a larger sample size, then species may come and go from the site. If you sample a larger area to get a bigger sample, you may no longer be sampling a single vegetation type. Such problems may be inherently indeterminate and may require a different approach.

Drawing conclusions

Care must be taken when drawing conclusions from a study. In the case of an empirical study, one needs to present both the predicted result and a measure of confidence. For the fertilizer example, if the result is +20% yield ±5%, then I will likely use it (assuming costs are right). If the result is +10% yield ±25%, I will not be too excited about it.

Sometimes people confuse the p value, which identifies how likely a result is due to chance, with the explanatory power. Studies in education and psychology are particularly prone to this confusion. If the result of a radical new education method is better at the 1% level but only accounts for 3% of the variance in the data, this is not a very strong result.

A particular caution is the experiment done in isolation. Medicines are often tested against a placebo, and may pass this test. However, what the patient wants to know is which medication is best. As another example, in studies of the species abundance distribution, discussed above, a common practice is to discuss a mathematical distribution and show that it fits the data, but without comparing it with the way other competing distributions fit the data. A fit of the data to the model does not prove the theory unless other postulated models can not fit the data. Reviewers should really not allow this. It is rarely the case that a single study either unambiguously proves or disproves anything. One can say that it lends support to a theory, or tests a particular prediction of a theory, but proof or disproof are generally cumulative over many studies. Many theories and results are never followed up on, and a particular discovery many need to be made several times before becoming part of textbook science.

The implications of a study are often exaggerated in one of several ways. This is particularly so in the press releases and news items about a piece of research. A common type of exaggeration is to say that the results of some biochemical or genetic study could lead to a cure for cancer. This type of statement has been made thousands of times as scientists try to justify their work, but most cancers still can not be cured. These studies are usually just small pieces in a very large and very difficult cancer research program. Similarly, studies are extrapolated to conclude that certain chemicals present huge dangers, when this result is beyond the implications of the study. Often authors draw conclusions about public policy from their work that are not necessarily inevitable. Just because the journal allows you to draw implications about your work does not mean you should. This is discussed further in the final section.

4.2   ETHICS

Doctors have a code of ethics and must keep their license in order to practice. Lawyers likewise have ethical guidelines and can be disbarred for violating them. Scientists do not have one and do not need a license to practice, though beauticians do. Why?

The Ph.D. is the common entry point for academic scientists but is not strictly necessary for doing scientific work or publishing. It is likewise common for scientists to publish in fields far from their official training. There is no license to lose and even amateurs can make contributions. This difference results because the scientific enterprise is fundamentally creative. A scientist devises experiments and conducts studies to try to add to the body of established scientific knowledge. None of this is guaranteed to be right until it is tested and challenged by others. Whereas doctors and lawyers have access to potentially damaging information on their clients, scientists are not in such a position. In this regard they are like artists.

This has led to the common misconception, however, that science as a practice is automatically free from ethical issues in the same sense that an artist need not worry about such things. Science in fact involves money, sometimes lots of money, and where there is money there is the possibility for misuse of that money. Scientists can also find themselves in positions of authority or influence. In such settings it is possible to be faced with conflicts of interest. No formal statement of how to deal with these issues is handed out with the Ph.D. diploma, so it is important to consider them here. The issues I will be discussing are not hypothetical. Many are encountered by every scientist and can seriously impact a person's career.

Treatment of students

In academia, successful scientists usually advise graduate students. This is so common that it is easy to overlook the inherent conflict of interest in their relationship. The professor gets the maximum out of the students if he or she treats them as cheap labor (cheaper by far than a technician). In this capacity the student should work on well-defined projects doing just what the professor wants, and should stick around as long as possible (e.g. the eight year Ph.D. program). The maximum benefits to the student are obtained by conducting his or her own work (with advice) and graduating as soon as possible. A common manifestation of this conflict of interest is disputes over credit for discoveries in a professor's lab. Some laboratory chiefs put their name first on any publications from their lab while others put their names last, as advisors. For others, it depends on the circumstance. The key here is for the professor to recognize that he or she has power and can easily take unfair advantage of the students. Students should similarly be on guard against assuming that their advisors will look after them in a disinterested way.

Reviewing

Almost every practicing scientist soon finds himself asked to review manuscripts and proposals. There are once again conflicts of interest. It is very easy, for instance, to give a little extra consideration to items produced by friends or to be extra critical of competitors. I have heard people insist that such things do not influence their views, but the safest course of action when the item is possibly open to personal bias is to simply decline the review.

In high school, my friends and I started a literary magazine. In discussing how to review the work, much of it likely to come from our own group, we immediately agreed on double-blind reviews, in which the names of the authors were redacted (I don't think we actually knew that word . . .). A few science journals have tried this, but most do not. Why? It seems that names and institutional affiliations help people fill in the undocumented aspects of the science, which helps them judge the work. For example, if a new technique has been recently developed at a particular lab, only scientists with affiliations there are likely to be proficient with the technique. Or, we may happen to know that Dr. Smith has done a lot of good work on a particular topic. These are valid considerations because expertise is not easily evaluated from the public documentation (i.e. scientific paper) of a project. Still, the temptation should be resisted to give prominent individuals a free pass.

A slightly different conflict of interest arises when reviewing proposals. The information in a proposal is in essence proprietary. The scientist is putting forward a plan of work detailing hypotheses or new results that will not be public for perhaps many years. It is easy to run off with these ideas, some of which might even have commercial value. Recognizing this risk, a funding agency (e.g. the U.S. National Science Foundation [NSF]) may ask reviewers to destroy proposals after they review them, but there is no enforcement of this. Besides, you can't make yourself forget what you have read! I have heard stories about proposal ideas being stolen almost verbatim, though this can be hard to prove. Thus if there is any chance that a proposal contains material that might be tempting, it is best to simply not review it.

For reviewing both proposals and manuscripts, another conflict of interest arises due to the fact that scientists provide their reviews free of charge. It is in the interest of the journal or funding agency for the scientist to do a thorough job, but the scientist should try to do the review as quickly as possible. The problem with conducting a quick review is that both type I and type II errors are easily committed. Falsely accepting a paper that should be rejected can occur because it seems OK but is really either not novel or has mistakes that are not obvious. Such papers take up journal space. Falsely rejecting papers that should be accepted can occur because the novelty is not recognized due to careless reading or because the reviewer did not understand it. Novel analyses can be difficult to assimilate and therefore may require extra time by the reviewer. I have a chemist friend who actually does experiments to test the work he is reviewing. This latter type of rejection is common and holds back the progress of the field. Part of the problem here is pride: the reviewer may be loath to admit that he does not understand the work, and instead of reading it again simply trashes it.

Small sins

There are many small ethical lapses that occur frequently in science. People do not lose their jobs over them, but that does not make them acceptable. The victims of these lapses are sometimes individuals whose careers are damaged and sometimes the general community of science whose trust is impaired.

One ethical issue is the undue claiming of credit for a discovery. This can of course occur when a person in a position of power claims credit for a discovery made in his lab or institute. This is a slippery slope issue. At one extreme, if the lab head writes the proposals, gets the grants, and supervises the work, then he probably deserves much of the credit. In other cases, however, the lab head has a more administrative role and his claiming of credit may begin to approach theft. Credit for a major discovery can be very important for the career of a junior scientist, and taking that credit unfairly can severely damage his or her future job prospects.

A related ethical lapse results from failure to acknowledge the contributions of prior studies to your results. Science is incremental and is a collective problem solving process. Almost always a particular result builds on past work as an improvement or modification. Because the literature is vast, it is easy to miss key literature related to a project. This is not really an ethical lapse, but sometimes it would be seemly to try a little harder to search out prior work. The ethical lapse comes when an author intentionally neglects to mention past work on which he or she built, thus making his or her results seem like a bigger breakthrough than they are. The same effect is obtained by giving overly brief or cursory references to past work.

More serious is when an author treats competing ideas unfairly in the introduction and discussion sections of a paper. It is too easy to disparage alternative theories, to give straw-man or distorted accounts which are then easily dismissed, or to only discuss selective results. If one's results or theory contradict existing work, it is imperative that prior work be described accurately when comparing it with one's new results. Ad hominem attacks on critics are never acceptable.

Finally, in some fields it is easy to repackage existing work using new terminology and make it look like a new discovery. This is common in psychology, for example. One can "rebrand" interpersonal theory as transactional theory or call a neurosis a coping mode deficit. Calling something by a new name will only fool some of your colleagues, and is not really a "discovery."

A related type of exaggeration is to stretch the significance of your results as a way to attract attention or funds. For instance, it is common for a university to release a claim that Dr. Smith's work on cell signaling (or something) will lead to a cure for cancer. Most research is incremental and only rarely does it rise to the level of "cure for cancer" or "clean energy source" or other such major breakthrough. Making such claims is a way to falsely claim credit for social benefits that have not yet happened and may not ever happen.

Humans are territorial about their homes and belongings, so it is not surprising that people get territorial about their knowledge domain. However, the world of science is wide open because knowledge is not open to ownership. Thus attempts to enforce a self-defined ownership of some domain of knowledge can only lead to inappropriate behavior. For example, as a postdoc I went for an interview for a faculty position. Even that early in my career I had a reputation for working on many different subjects within my field and for being prolific. During a meeting with one of the faculty members, he was telling me about his long-term research on a single topic and then he looked right at me and said, "of course, I don't want you just waltzing in here and solving this problem." I'm pretty sure I didn't get his vote. But, how strange! He was warning me not to work on a problem, as if he could keep it to himself. I have heard similar stories from others. In one case a professor offered a postdoc (not his own) a data set on the genetics of species B if she would stop working on species A, which was "his" domain. At least he was bargaining, but really this is unacceptable.

Grants and contracts are a problematic area because money is involved. When one applies for a grant, there is a description of the type of research that the granting agency seeks to have done. This may range from quite basic for NSF, to quite applied for the military or a corporation. For example, a company may look for safety tests of a product or toxicity tests for a chemical. In response to their request for proposal (RFP), the applicant writes a proposal. This proposal contains background information, a description of the work to be performed, and a budget. The review panel for the proposal (or the funding officer) will decide which proposals to fund based on the description of the work, but also on the reputation of the investigators on the proposal and the reputation of the institutions where the work will be done. There are various forms of trickery with grants that scientists do that are really unethical.

A common trick is to ask for money for work that has already been largely done. In the proposal, only "preliminary" data is shown, which is partial data from the work completed. In this way, the scientist knows what the results are and can speak with high confidence about how the study will turn out. While this does overcome the tendency for proposal reviewers to demand proof that is really not possible to give, it violates the spirit of the grant process.

In some situations, it is possible to be a little vague about who will do what work when only a general budget is required. This can lead to various abuses. It may look, for example, as if the senior scientist will be doing the work but in fact students will be doing most of it. In other cases with multiple investigators listed, some of them may get dropped when the money arrives. In both cases the granting entity is being deceived. In the latter case, those dropped have had both their time and intellectual property stolen. I have witnessed all of these situations personally.

Fraud

Fraud is a legal term referring to theft by deception. For example, one who sells land that is in a swamp using a photo of some other place is guilty of fraud. Since scientific work is generally not being sold, it is being published, and the author may even have to pay to have it published (quaintly called "page charges"), how can there be fraud? In science, the term is used in a slightly mutated sense to refer to deception, and in particular to fabrication. It is considered the worst crime in science and can cost someone his or her job. We can understand the aversion of scientists to fraud by noting how it undermines trust within the community of scientists. Fraud as a concept, however, is not without complications.

Episodes of fraud occur sometimes in high profile areas of science, such as human cloning, HIV/AIDS, and medicine. In such areas, there is a rapid pace of discovery and extreme pressure to obtain results. As well, the reward for success can be great. Thus in certain psychology and medical studies patients were recorded who did not exist. Sometimes bad data (which does not support the researcher's pet theory) is deleted. Much fraud is probably not detected because it is simply overlooked.

A complication with detecting fraud is that sometimes a scientist is entirely convinced of his or her theory and data, but is deluded. Examples include N-rays, polywater, cold fusion, and para-psychology. Under such delusions, it is easy to select data that support the theory and explain away failures. Note, however, that in such cases we have largely self-deception, and the scientists themselves may be honest (i.e. are not fabricating data). Such cases differ only in degree from the common experiences of scientists who become passionate about a theory that may turn out to be false. Similarly, anyone who has worked with data will have had the experience of testing some relationship and getting a result that simply did not make sense and was therefore thrown out. Is this deception? No. Science does not consist of experiments that may have been done poorly and incoherent plots of x versus y. Those relationships that seem to be explanatory or predictive survive.

This means that it is not so obvious when fraud has been committed. It is not a crime like a financial crime that can be investigated by detectives. Instead, another cure is in general in order.

The cure for science fraud is not laws or detectives but transparency. It is characteristic of most fraud that something is hidden from the reader. If a piece of work is important, others will test it, build on it, and elaborate it. Mistakes are often found when attempts to replicate a study fail. For example, if a small sample study says broccoli is good for you but other studies are unable to verify these results, then the first study was probably a statistical fluke or based on a nonrepresentative population. In fraudulent work the methods and statistics are often vaguely described. Increased clarity in studies can be achieved by using an appendix and/or online supplemental information to provide the extra details that other scientists need to verify the work. This should include computer programs, data, statistical package runs, mathematical derivations, maps, and photographs. Journals that do not provide online archives should be encouraged to do so.

Cherry picking

It is very tempting to gather evidence that supports one's pet theory. This cherry picking of evidence can occur at the data gathering phase quite easily in certain fields. For example, cases in which a particular psychological treatment proved efficacious will stand out. Reporting only on such successes is a common practice in popular psychology books (e.g. on relationships or marriage). There is no mention of the patients who stopped therapy because they thought the therapist was an idiot or who blame him for their divorce. This is why clinical trials are so important in medicine.

Cherry picking can occur during data analysis. If many different relationships are tested for in a large data set it is likely that some of them will be significant purely by chance. Reporting only the positive results is deceptive. There are methods for correcting for the number of tests performed (e.g. Bonferroni's correction) and these should be used. Conversely, one should perform and report adverse results. For example, a test for randomness that can not be rejected should be reported.

It is also not appropriate to treat nonsignificant results (e.g. $p > 0.05$) as still being a result.

Selective picking of results can be used to support a particular point of view. For example, if someone is looking for environmental impacts from some practice or activity, it is easy to sample lots of things (species, water quality, metrics) and only report those with an impact. One should at the least report that 15 things were tested but only one was impacted.

In the discussion section, one is supposed to compare one's new results, method, or theory with past work. One should compare with work that agrees with your new results and work that does not. To only pick studies for comparison that agree with yours is cherry picking and will not fool reviewers who may return their reviews with a long list of papers you "missed."

While Machiavellian scientists can sometimes get ahead, they will not be admired and may even be lampooned by others. It is also a dangerous game to play. I have known scientists who cut corners and took advantage and bullied and things have turned out badly for them. High road or low road, it is your choice as a scientist. Given the autonomy most scientists have, it will be difficult to blame others when things go south due to ethical lapses.

4.3   PUBLISHING

Publishing is a critical part of the job of a scientist. Science is a public endeavor, a body of knowledge. Thus contributions to science must be public to be counted. Giving a talk about some discovery can be exciting but it is not enduring and reaches few people. It is helpful to approach publishing with a strategy in order to minimize headaches and maximize output. This section addresses this topic.

Types of publications

There are thousands of journals. These all differ in the format required, topics they allow, permissible length, etc. Some allow commentaries and reviews, whereas others do not. Some insist on first person narrative ("we collected samples") whereas others forbid it ("samples were collected"). It is thus critical to be aware of the journals in your field and their requirements.

There are several types of articles suitable for a journal. A research note is the easiest to get published. This type of paper addresses a single phenomenon and is only a few pages in the journal. It describes the results of a relatively discrete study, usually empirical. A standard research paper is rather longer. Review papers are often even longer but can only be submitted to certain journals. Reviews can be difficult to get published but can also be widely read. Large-scale studies written up as a monograph are another category. These can be difficult to publish due to their size. It can also be difficult to find reviewers for them.

Another way to look at a publication is in terms of topic. Empirical studies are an obvious category. These can either involve field data collection or experimentation. Methods papers are also common and cover topics such as methods for synthesis of some compound, methods for tagging animals and tracking them, etc. Statistical methods papers are an important category. Theoretical papers are also important but can be classified by reviewers as "just" a review paper, or worse as an opinion paper, and thus rejected. In a theoretical paper it is important to demonstrate that the analysis performed is substantial and leads to new knowledge.

Books are an important outlet for scientific knowledge. Books are of multiple types. A textbook summarizes established knowledge in some area. Introductory texts can be a significant source of income for classes taken widely such as calculus. However, the broader the market the more textbooks will be on offer. Advanced level and graduate texts will not bring much income, but can establish the reputation of the author. Specialized texts can provide an outlet for more complete exposition of some topic and again can help cement one's reputation. If a field is fast paced, it can be difficult to finish such a book before it is out of date. A sabbatical can be useful for this purpose. Sometimes books are written with each chapter written by different authors. This can alleviate the problem of taking so long to write the book that it is not timely, but often one or more authors are really late delivering their chapter. Finally, books that are popular can be a good way to communicate with a broader audience on topics from biodiversity to evolution to heart disease. It may be that such a work is not taken entirely seriously by all of one's colleagues.

Some institutions have outlets for publishing their own work. For example, the U.S. Forest Service has publications for its scientists. This allows work to be published, such as silvicultural guidelines or manuals for conducting a controlled burn that would not find an easy outlet in a journal. Similarly, the agriculture department publishes guidelines for farmers. Again, one's colleagues may look down on these types of publications, so one is advised not to focus on them exclusively. Some universities run their own book publishing houses, which may particularly favor their own authors.

A new type of publication is the software package. A software tool for dealing with animal survey data, for example, will enable users to more easily and reliably analyze their field data. Such tools are widely used and range from very simple to sophisticated. In fields like statistics, many of the tools have become commercial packages.

Symposium proceedings also deserve mention. This can seem like an easy way to get published: you give a talk and your paper goes in the proceedings without any of those pesky reviewers bothering you. However, proceedings are typically not indexed in search services and are not available for people to read. Their reputation is that preliminary and less polished work fills them up. They are thus a poor choice as an outlet for one's work.

Publish and/or perish

"Publish or perish" has long been a buzz-phrase in the world of science. Obtaining grants, getting tenure, and getting a job can all depend on publications. How they are treated, however, is far from consistent. A colleague was told by his department chair that his book essentially counted as no more than a single publication. Another was told that his research was great, but he could not get tenure until he published a book. In some cases one may be told that publishing is critical, but publishing more papers does not in fact lead to a higher salary or more rapid promotion. That is, the treatment of publications can be idiosyncratic and inconsistent.

A second tricky thing is authorship practices. If all your work is written by yourself and published alone (sole author), this would seem to indicate that you are competent to carry out the work. It can perversely be viewed as showing that you don't collaborate well. On the other hand, a professor with a lab full of students and postdocs can sometimes crank out the papers based largely on their work. This may seem like a good thing, but sometimes the word gets out that Professor X is just putting his name on papers his students did. This can damage a reputation.

Publishing can damage a career in other ways, as well. Those tempted to cut corners, to publish the same data multiple times, or to exaggerate the importance of their results can be found out, with devastating consequences. Even an honest scientist can become overly excited about preliminary results and rush them into print. When later work shows this result to be invalid, it can be detrimental.

Timing

In scientific publishing, timing is critical. Great importance is attached to priority of discovery. Only one person or team gets to be first to discover a new species or elementary particle. Later papers showing the same thing will be rejected unless specifically couched as confirming or refining the original discovery. There is a tradeoff, of course, because longer study of a problem is likely to yield more definitive results. In general, it is safer to spend more time on a topic that is specific (tigers versus laws of physics), particularly if few are working on this topic or it is place specific.

When a new technique or device is developed, it can be a great opportunity to be "first." For example, the first paper to use fractals for characterizing branching networks or the first to use the electron microscope or isotopic traces will probably receive a lot of interest. New techniques and tools quickly become standard, so it is critical to be an early adopter if the goal is to get credit for being first.

Timing can be critical for a synthesis or review paper. Following a burst of activity on some topic, a body of literature may quickly accumulate that is unconsolidated and may appear contradictory or that may need to be related to theory. The timing for writing a synthesis piece is critical. After others have written one, it will be harder to get one published.

There are fads in science. After all, there is no inherent reason to study many of the topics that people study (besides the ones that may make money for us, of course). Why indeed study supernovas or whales except that they are awesome? So, fads come and go. For a while, fractals are hot, or bucky balls, or mesocosm studies, or superconductors. Fads may end because the topic is exhausted, because unforeseen difficulties take the fun out of it (e.g. superconductors), because the topic becomes mainstream (e.g. fiber optics, gene sequencing), or because it doesn't lead to the cure for cancer that was promised. In any case, if you do fall in with a fad (which is OK), just remember that the party may be short. If you spend four years studying something that is a short-lived fad, it will seem uninteresting to reviewers and readers by the time you are done.

Mechanics of writing

Writing a scientific paper is not easy. It is not the format, per se, but the difficulty of conveying something complex in a very limited space. This is especially so because it is not the case that all readers (or even reviewers) can be expected to have the same degree of expertise or type of training. If you assume they all understand spectral analysis or Kalman filters, you will be criticized for jargon and not being clear enough. If you explain such "elementary" concepts at length, you are being wordy and exceeding space limitations. It's a tough crowd.

There are some concrete steps that can be taken. Practice helps. Second, revise, revise, revise. If each revision makes the paper slightly better, and slightly more precise, then eventually it may become quite good. During the course of making revisions, it is critical to put the paper away for a week so that you can view it with fresh eyes. Another trick is to ask oneself how your audience will read it. Perhaps you are assuming too much in terms of their background, so that certain things need to be stated more explicitly or even defined. It is also helpful to keep a list of unresolved issues (e.g. what about outliers? did I control all variables?) so that you can ponder them in the background and not let them slip by.

Getting friendly reviews can be enormously helpful. Your colleagues may be aware of papers you neglected to cite or perhaps will see a mistake in your statistics. Be sure to acknowledge them.

Controversy

There is always the potential for controversy in science. This can arise from multiple sources. There are fundamental disagreements within science about such issues as valid statistical approaches, where different individuals may adopt a Bayesian, frequentist, or information theoretic approach. Theories can be ambiguous as to what they predict. Results of an experiment can sometimes be viewed in different ways, in which case making a strong interpretation of them is just asking for debate. Many times a new result may suggest that past studies suffered from small sample sizes, confounded experimental designs, failure to control for some factor, or some other flaw. It is best not to poke prior researchers in the eye in your paper by pointing out how flawed their work was. It is instead better to describe your study as more comprehensive or as taking a different approach.

There are topics that are inherently controversial. In such cases, others may decide they don't like your results regardless of your methods. For a long time it was not acceptable to suggest that disease might operate differently in men and women or in different ethnic groups, because this was taken to imply discrimination, and thus clinical studies tended to be done with white men. It has gradually become clear that risk factors differ by group and responses to treatments such as medicines can vary radically. Thus studies have begun to consider gender and ethnicity, though danger is still present if you address this topic. Other controversial topics include why boys fall behind in school but dominate math, science, and engineering; educational methods; HIV transmission studies; anything to do with race; risks from pesticides or other chemicals; climate change; child rearing effects; and many others. When writing about any such inherently controversial topic, it is advisable to give extra attention to wording, to clarifying exactly what can and can not be concluded from the data at hand, and to avoid extrapolation of the results. That is, from a study with 50 people on diet, one should not recommend an overhaul of school lunch programs. Even with such care, one may receive negative press and criticism.

In some cases, a study will necessarily challenge past work because it presents a new theory or more conclusive data. In such a case, do not expect those whose work is shown to be wrong to take it well. Comments or critiques may be written about your work, which can be nasty, personal, and even quite strange. Editors seem to allow bad behavior in a comment on a study, though they should not. In most cases, if your work is commented on you are supposed to be given a chance to reply, but this is not always done. If you do get to write a reply, do not stoop as low as your critics.

Journal reviews

In order to get anything published in a journal or even a book, the material must first pass peer review. This very imperfect process is at once beneficial and annoying. It is usually capable of weeding out the real dreck or incomplete work, but can also lead to quality work being rejected. For example, the reviewer may fail to understand what you did due to careless reading or lack of appropriate background. His or her recommendation for rejection is usually considered valid by the editor, though I don't know what happened to the editor's own judgment in such cases. Comments can be truly bizarre. In some cases it is not even possible to understand what the reviewer meant.

For cases where you are given a chance to rewrite the paper and resubmit it, there are three types of comments to deal with, and all three must be answered in detail when you send it back. The first case is the set of valid criticisms, which may suggest a different statistical test or mention important references you missed. The response here is simple: make the changes and thank this reviewer. The second type is the nit-picking review. This reviewer focuses on little stuff like formats. Again this is easy, just correct the little stuff and you are done. The third type of reviewer is the obnoxious one who missed the point entirely, simply doesn't believe you, or wants you to redo the study in an entirely different way, perhaps with an impossibly large sample size. The review may actually be gibberish. Your reply to these comments must be patient and thorough. You can't say that he missed the point and must have read it while watching TV. Editors should reject such flaky reviews, but often don't.

Conclusion

Overall, writing is a critical part of being a scientist. Incoherent writing causes many papers to be rejected or not read once published. Conversely, top-notch writing has led to many scientists being widely cited even though not so original (or even correct). Sad but true. Writing style per se is not covered further here except to note that one should not look to writing guides produced by English teachers for guidance, because scientific writing is entirely different from such writing and is not "literary."

4.4 SCIENCE INFORMING POLICY: THE MYTH OF OBJECTIVE SCIENCE ADVICE

There are an increasing number of issues where public policy hinges on scientific advice. From public health to pollution control to stem cell research to climate change, highly technical aspects of the science may be central to understanding and solving a problem. Naturally, experts are often consulted to help inform management or policy making. Bringing experts in does not, however, automatically lead to clarity. The problem is that while science as a whole is concerned with how the world actually works and is autonomous from individual wishes and points of view, individual scientists are not free from subjectivity, bias, and agendas, as much as they might try to be and as much as we might wish them to be (e.g. Lackey, 2007).

In science, individuals are not vested with authority to rule on scientific questions. That is, no group or class holds a higher, privileged position. Rather, every theory and model is, or should be, under constant assault by new methods, data, and theories by anyone at any time.

Einstein overthrew Newton, for example, even though an unknown at the time. Note, however, that he could never accept quantum theory. His fame and brilliance were no protection against being wrong, nor is any expert guaranteed to be right. At best, they will *often* be right, but even then only partially so. This incomplete omnicompetence (and objectivity) results both from the limitations of science itself and from biases inherent in the human psyche. In this section I explore this idea of irreducible bias and then provide a solution that reflects the actual practice of scientific research, but not of current policy advice practices.

In the interest of disclosing my own biases, I am a research scientist working for an industry nonprofit research corporation. I have served on many panels and committees advising government, industry, and nonprofits, as well as review panels of various sorts.

Bias and the practice of science

The problem with ancient philosophy was that it was too easy to believe what people preferred to believe or what made a good story. Modern science is able to make progress by imposing hurdles that reduce our ability to fool ourselves and others. Experimental design and statistics were developed to counter the tendency to see what we want to see and build theoretical castles on foundations of a few data points. In medicine, the tendency of patients to believe they will get better and thus to actually get better (the placebo effect), or to merely report that they are better, has led to the double blind clinical trial. The ambiguity inherent in natural history observations led to the development of controlled experiments. We are fond of our own data and theories, and thus peer review has become a critical tool for reducing the tendency to be less than critical of our own work. Replication of a study helps to guarantee that some subtle bias or error did not lead to the reported results.

All of these practices help reduce self-delusion and increase (not guarantee) objectivity. Nevertheless, except when basic scientific facts are involved, the ability of individual scientists to reason clearly about complex issues and give objective advice is seriously constrained both by psychological factors and by limitations of the scientific process itself. It is also worth noting that in many areas of science that impinge on public policy it is difficult or impossible to implement some of these safeguards, which makes the problems I will be discussing worse. For example, policies designed to slow AIDS transmission can not be tested experimentally, for ethical reasons. The spread of avian influenza or severe acute respiratory syndrome (SARS) can only be modeled and reasoned about. There is no way to prove the actual risk of nuclear reactors except to build lots of them and see what happens. We are forced to reason about the extinction of the dinosaurs, since we lack a time machine. In all such cases, human reasoning, inferential statistics, and models must be used instead of controlled experiments, and all of these are less than perfect ways of obtaining information.

One of the complications when science must be applied to management and policy questions is that science is misconstrued to be about facts, which are not open to question, when this is really only the case for simple questions. For example, if the question involves the calculation of distances on the globe, then there are exact answers and all other answers are wrong. But if the question is "Is travel to Mars feasible?" or "What diet will reduce cancer risk?", then the answers are complicated and are never guaranteed to be complete. Those who took a few sciences classes many years ago may believe in the "science = facts" view, whereas those who conduct research should know better. Even here, it is often observed that scientists treat the conclusions of their research as "facts," although they are inferences.

How science actually works: theory and knowledge bias

The scientific process is supposed to work by a progression of discrete steps. A critical experiment or data should cause a bad theory to be rejected and a new, better one to be accepted. I say "should" because this is not what actually happens. The way science actually works impinges on the ability to give objective advice.

In any field, it is possible for a major theory to be in a state of ambiguity for long periods of time. In this situation, alternate camps will refer to their own view of the issue when giving advice, and may do so with the appearance of certainty. This ambiguous or limbo state can occur because a theory is difficult to test. For example, for decades there have been alternative theories for the origin of modern humans: out of Africa, simultaneous evolution over multiple centers, and most recently, out of the Near East. Fossils are only slowly discovered and are themselves ambiguous, so that testing the alternatives is very difficult. In physics, any proposed unified theory is both mathematically difficult and requires enormous energies in an accelerator to test its predictions (if at all). In such cases, advocates of competing theories may feel justified in supporting a particular view, but really none are unequivocally supported by existing data. In such cases, strong but opposite opinions will be offered by advocates for different positions. These differing opinions can be put in context and properly balanced only by considering the current unresolved state of the science. That is, in such cases there simply is *not* a correct answer that can be obtained from experts because they fundamentally disagree with one another. This is a theory bias.

Logically, when new data are obtained that fundamentally overturn an old concept, everyone should snap to attention and change their opinions. Sometimes this happens, but often it does not. It took decades for quantum theory and continental drift to be fully accepted. Why? There are two fundamental reasons. First, it takes time to come to grips with any major scientific change. Scientists are busy with their own work and if the new developments are not right in their topic area, it may take quite some time to notice and master the new ideas, if ever. Some may find the new material too difficult and simply avoid it. Second, some people (even scientists) resist change. They may have come up through school with the old theory and feel comfortable with it, and thus not be willing to adjust to the new. Those who helped develop the old theory are particularly unlikely to adopt the new theory quickly, because of pride, sunk costs, and difficulty reconceptualizing their research program.

For these reasons, it is rare for any individual scientist to be completely up to date on all aspects of his or her field. It is not uncommon to see papers (at least before peer review!) neglect the latest developments and cite work that has been superseded. Scientists in nonresearch positions may fall particularly far behind. The result is knowledge bias, where randomly chosen experts may give out-of-date advice, in spite of them all having "adequate" credentials. Experts also can be unaware of when (or on what) they are out of date.

Advocacy bias

At the level of the individual scientist, the standard practice of science is full of advocacy. When conducting basic science, a researcher does a study, develops a theory, or develops a new method. This work is not presented objectively in the scientific paper, though it is rather dressed up that way. The author makes every effort to present the work in the best light. He shows how it builds on or supersedes past studies by pointing out the flaws in past work (what they can't explain, artifacts they create, lack of fit to data) and how those flaws are overcome by the new work. He also tries to show how important the work is by discussing its implications. As a counterbalance, reviewers may make the author give a more balanced presentation of past work and substantiate his claims for superiority and importance. As a further balance, other authors will point out the flaws in the new work that the author either did not perceive or happened to neglect to mention. It is a rare scientist who presents a balanced view of his or her own work. This type of advocacy bias is implicitly understood by the scientific community, though perhaps not by the public, including the consumers of scientific information.

When a scientific question involves applied aspects (public health, medical research, agriculture, etc.) the problem of advocacy bias becomes greater. Now when the scientist has a pet theory or result, it is not merely about something abstract such as photosynthesis, but about what foods people should eat, or how farmers should farm. Usually, a study deals with only part of an applied issue. For example, a study might show how forest patch size affects an endangered species. Implications are then drawn for managing forests to protect that species. The study did not include other aspects of the problem, such as economics or natural disturbance, and thus the suggested management activities are an inference or extrapolation and not something that inevitably follows from the analysis. Separating what are the real implications of the research from the scientist's personal views is not always easy.

It is very difficult to simultaneously hold an advocacy position and give objective scientific advice (Lackey, 2007). It is easy for one's personal worldview or politics to make it seem "obvious" what the best solution might be, but others may come to a very different "obvious" conclusion. For example, different social scientists have come to the conclusion that poverty is best "cured" by stimulating the overall economy, raising the minimum wage, creating or getting rid of low income housing, starting or stopping welfare, keeping families together, improving schools, etc. All of these are "obvious" to those proposing them, but some are not so easy to do (how do you keep families together?) or have unproven efficacy. It is equally easy to believe that one's own views are correct ("objective") but that those who disagree with you are advocates.

Political advocacy is well-enough known that people can be on guard about it, but there are other more subtle forms of advocacy that also affect expert advice. One type is the toolkit bias. If someone has become an expert on a certain set of tools, they may advocate for the use of those tools whether they are the best for the job or not. A modeler may advocate for building a model to solve every problem. Surgeons will recommend surgical solutions to medical problems. Taxonomists recommend surveys. For certain conservationists, a nature reserve is the solution to every conservation problem. A similar problem arises from those who push their favorite theory.

There are thus several types of advocacy bias. Authors always try to put their own work in the best light when they publish or give talks. If they did not, it would be ignored and not even get into print. Even unconscious advocacy makes one's preferred solution seem obvious, but others with different views will come to different "obvious" recommendations.

Normative bias

In the context here, a norm is a goal value to which the expert subscribes, often unconsciously. Some of the norms of scientists involve the types of scientific problems to which they are attracted. For example, scientists are attracted to "interesting" problems (often idiosyncratically defined) as well as "important" problems (which will get lots of attention). This means that rare things (e.g. rare species) have more inherent interest (and thus value) to scientists than perhaps to the general public. A common norm is that naturalness has greater value. Another norm is that complex natural things tend to be more aesthetically pleasing (e.g. Yosemite) and thus more valuable. These norms influence policy. For example, in Western dry forests fire is a natural disturbance, the absence of which is itself a disturbance that increases the risk of catastrophic fire. Even though this is pretty widely understood, there is still resistance to managing these forests because management violates the "naturalness" norm. Scientists often treat these norms as if they were objective facts (i.e. follow inevitably from the science) when they really are not (Lackey, 2007).

Disciplinary bias

It is natural to be more familiar with topics in one's own discipline. One's field of study also comes with a particular suite of analytic tools and conceptual frameworks. The need to focus, critical to conducting research, can lead to blind spots when it comes to interfacing disciplinary knowledge with other disciplines or social problems. For example, in a "multiple use" forest, it is easy for the fisheries expert to give recommendations solely related to fish, the wildlife expert to suggest how deer habitat can be enhanced, and so on, with none of them clear on how all these requirements interact nor what they cost. When my father had a heart attack and then a stroke, I witnessed a debate between the neurologist and the cardiologist over which blood-thinning medicine to give him. Each favored the medicine that posed the lesser risk to the part of the body in which they specialized. We were unable to get a risk assessment that was objective.

There are other disciplinary biases as well. Theoreticians tend to be overly fond of their models, and may view them as more real than mere data. Policy recommendations based on a model may not be implementable in the real world. In contrast, strict empiricists may be unwilling to extrapolate beyond their experiments. Lumpers and split-ters also differ in their advice. A taxonomist may view every variant of a species as unique whereas an ecologist may view them as functionally equivalent, with obviously different conservation implications.

The tools and approaches used within particular disciplines have limitations, not all of which are obvious to those accustomed to using them and most of which are not obvious to outsiders. These limitations will bias the advice an expert gives. For example, when ocean surveys are based on net samples, soft-bodied creatures (e.g. jellyfish) are sys-tematically undersampled because they are destroyed by the capture process. Even when oceanographers are aware of this bias, others who use their surveys may not be aware of it. When economists forecast the effects of changing the minimum wage or interest rates, their forecasts are based on various assumptions and simplifications which limit the certainty of the projections. Thus economic forecasts are never precise and "surprises" occur regularly, such as the unexpected subprime home lending meltdown of 2007 or stagflation some decades ago. Consumers of such forecasts need to be aware that such surprises are common, and are thus not "surprising" at all.

It is not merely that specialists in different disciplines have biases, they have almost irreducible biases, of which they are largely unaware and from which they can not escape. If Bill were not so obsessed with turtles, he would not be any good doing research on them, but this very strength may make him blind to other facets of a problem. Biologists often have no training in economics, for example. Because the issues on which expert advice is sought often involve the intersection of topics and disciplines, it is critical that disciplinary bias be kept in mind.

Omnicompetence bias

Related to the disciplinary bias problem is the omnicompetence bias problem. Scientists who are successful and are often called on as experts may begin to view their advice as always useful, even if it is outside their area of training. We can observe this with Nobel Prize winners who may opine on topics that have nothing to do with their training, such as third world development or economics or peace efforts. At this point they move from expert to pundit (one who has interesting opinions). This is a bias because what one gets is a personal opinion but dressed up in the clothing of an expert. I have personally seen such experts talk nonsense when they venture outside their field and into mine, but with perfect confidence.

Psychological bias

People obviously differ. In theory this should not affect the practice of science, but it clearly does. Someone who is risk averse will avoid speculative topics and new theories and may also be prone to see disaster looming. Someone intolerant of ambiguity will prefer clearly defined research projects and may simply refuse to engage with messy policy issues. Certain people are fond of grand theories and fall in love with them, whereas others like puzzles. Human reasoning is not highly logical (like Spock on *Star Trek*). It is easy to come to conclusions that are not logically implied by the data or that are internally contradictory. The same people who say that saving for retirement is important may be spending beyond their means. Pragmatists and idealists differ in the way they view problems. The idealist focuses on how things should be or the best solution. This is often a spur to change. On the other hand, the pragmatist focuses on what can be done with the available data or funds, and this type of solution is more likely to be feasible, though it could also represent a bandaid fix when a complete rework is needed (e.g. patching a street that needs to be repaved). In technical fields, many people are detail oriented. If you are going to do experiments, the details matter. But in many cases expert advice concerns the big picture, not just the details, and detail types and big picture types will give very different advice. Thus, psychological factors can lead to various types of bias.

Values and bias

Science is supposed to be value free, to be about what *is* rather than what we wish were so. When interpreting scientific results or conveying them to the public, however, it can be very hard to separate our values from the science itself. Let us say that some environmental effect is assumed to be bad, and our study confirms that this bad thing is harmful. It is very tempting to announce categorically that this thing is bad. This leap from statistical significance to policy importance may not be justified. For example, the harm may be so rare or so small that one could never detect it outside the lab. It can also be that the harmful thing also has beneficial effects. For example, preservatives in foods can not really be said to be without any harm, but at the same time they prevent many people from getting sick and even dying and reduce food spoilage and thus save money. As another example, too much sun causes skin cancer, obviously. It is becoming clear, however, that in northern regions people can get too little sun. The lack of vitamin D can then cause a variety of health issues. It is thus important not to let a categorical label of "harmful" override questions of benefits or magnitudes of both benefits and harm. This is especially so when claims are made about some harm based on weak or no evidence at all (cell phones, Alar used by fruit growers, children playing with toy soldiers, etc.).

Values can intrude on basic theory in science (e.g. Kellow, 2007). This is less true in the physical sciences, but is pervasive in medicine, anthropology, sociology, psychology, ecology, and certain other fields. In ecology, for example, the positive regard for nature and its beauty has led to strongly held views that nature is in balance, in equilibrium, in maximal order, tightly coevolved, and so on. Over the decades many of these virtue concepts have been confronted with contradictory empirical evidence, but they have tended to linger. In psychology, feminist ideology insisted that the sexes only differed due to societal expectations and this led to efforts to ignore or deny that even baby boys and girls show differences apart from cultural influences. Anthropology has had trouble dealing dispassionately with evidence for human migrations, evidence for or against warfare in past societies, interpretations of past religions, and many other issues because assumptions about what is possible or normative so easily color interpretation of bones and ruins. In medicine, preconceived ideas of what is healthy or good for you have made it difficult to study certain issues.

Conclusions

Science as a process does not reside within single individuals, but in the interplay of individuals. Individual scientists can be stubborn and refuse to change their minds. They can fall in love with their own ideas and refuse to give them up. Various subtle and not so subtle biases afflict their work. This is even more so when asked for their advice, because then they are even more tempted to become advocates. I propose two Laws of Subjectivity: (1) It is very difficult to be aware of one's own biases or blind spots; and (2) Any attempt to point out bias in others will be taken as an ad hominem attack, rather than a valid criticism.

The checks and balances of science play out over years and decades and depend on constant debate. Does a theory stand up to experimental test? Can a study be replicated? Does this method produce artifacts? At any given moment much of it is in flux and not settled. Some will see the current evidence as compelling for a given viewpoint, but others will not.

The moment an expert makes a pronouncement about heart disease or pollution or airline safety as if his word were final, we should become concerned. No individual scientist can make authoritative statements nor tell society what it *must* do. Scientists are not the only smart people around. Even smart people can pronounce the *Titanic* unsinkable. Even Einstein made mistakes. On any nontrivial issue, there are always multiple ways of slicing it and multiple possible solutions. Not all proposed solutions are efficacious or feasible. Any time someone insists we *must* listen to them it is clear that this is advocacy and not objective science.

The biases that afflict experts can be anticipated, both in general and in specific cases. This means that they can be corrected for or balanced by keeping the following points in mind. First, no matter how many scientists agree, it only takes one person to prove them wrong (remember the discovery of continental drift or the proof that ulcers are really caused by microbes rather than stress). Thus it pays to keep an ear out for competing views. Second, the state of any particular science can be one of competing or untested theories for literally decades. Just because we want a clear answer does not mean that one is available. This means that we must learn to tolerate ambiguity. Third, scientists flock with like-minded colleagues, forming cliques as bad as those in high school. If you ask advice of ten scientists who publish with each other, you are not getting ten opinions. Fourth, simple answers are almost certainly wrong. Fifth, if an expert is shouting loudly that you *must* listen to him, he is advocating a position, even if he denies it. Sixth, both the expert and the public need to remember that there are many subtle biases that can affect the viewpoints expressed by scientists, and that their advice is not truly objective.

To deal with bias and advocacy, it must be remembered that one can not trust any individual publication or expert to be right on any nontrivial topic. The Law of Unintended Consequences must be kept in mind. If an expert offers a guaranteed solution to some problem, with no downside or risks, it is probably a biased view. If the expert treats inferences as facts and makes claims to authority, one needs to do some double-checking of the advice. The consumer of expert advice clearly needs to exercise his or her critical faculties.

It is not just the consumers of expert advice that need to be alert, however. Those offering advice need a little introspection as well. It has become fashionable to include as part of the discussion section of a paper some application or broader significance of the work being presented. Recommendations may be made about public health or saving endangered species or forest management. Writers should per-haps ask themselves if this section goes beyond their expertise or reflects their personal biases. Is the author, an applied mathematician, really competent to identify practical fisheries management options and policies just because he or she has a nice model? Is the most pressing problem really to survey for endangered species before they are lost, or is this merely the author's favorite activity? Just because it is easy to give advice does not mean that it should be done casually.

For the working scientist, understanding bias can also make his or her work better. Any time one is on a panel or committee it can be confusing if one is not aware of the types of bias that others, and oneself, can exhibit. Understanding these biases can lead to better communication and help head off fruitless debate. Reviewers need to keep this in mind as well. If a paper is written to answer an empirical question it is not fair as a reviewer for you to demand that they address basic theory. Conversely, as an author you will get comments from reviewers that only make sense in the context of some sort of point of view. While writing a paper, it can be useful to ask oneself how this work might be perceived by people with different expertise or biases. In this way you can perhaps address their concerns up front and prevent rejection.

5

Summary: the strategic approach to science

Science as a career is fascinating and fulfilling. The popular picture of what scientists do, however, does not provide a useful guide to those considering a science career. The first section of this book provided an overview of science as a profession and what scientists do. The central focus of the book, however, is on the inner game of science. How do scientists solve problems and discover new things? How can they increase productivity? What techniques can they use? To address this issue, a strategic problem solving approach is proposed.

Strategic thinking is an overall technology for problem solving with the potential for increasing both the quantity and quality of work in fields where creation of novel solutions is critical, such as design, invention, institutional innovation, research, software development, and engineering. It is based on an understanding of both outward processes (actions) and inner operations (cognition, attitudes). It takes particular care to be alert to the inherent flaws in all aspects of the problem solving process, and builds in checks and balances against these flaws.

There are four factors involved in the successful development of an innovative creation: creative mental functioning, an understanding of the discovery process, an appreciation of the structure of multistep problems, and the utilization of reality checks. Creative mental functioning is necessary to be able to recognize problems, overcome intimidation by the status quo, overcome barriers to thought, and avoid stagnation. Various attitudes and work habits such as topic surfing and overcoming writer's block, contemplative walking, and not becoming an expert may allow one to increase innovative capacity and to recognize good ideas when they come along. Creative functioning alone, however, only produces bright ideas and not finished products. Discovery is the next key step that represents the development of the germ of an innovative idea. The key point in the context of strategic thinking is the recognition that discoveries are almost always initially amorphous and in need of elaboration before they can be communicated, tested, or formally documented. Elaboration is a fundamental but usually overlooked part of the innovation process. During the elaboration of an idea or invention, prototypes or models must be built, definitions or metrics developed, vague ideas made more concrete, and the idea must be communicated. It is necessary next to keep in mind that there are always opportunities for mistakes which may destroy a new discovery or bright idea, or which may require reworking. It was shown by a systems analysis of the problem solving process that multistep problems are particularly susceptible to the effects of mistakes, which waste effort in an exponential manner with increasing problem complexity. An understanding of the structure of multistep problems can help one avoid catastrophic waste of time, particularly by looking for ways to convert sequential into iterative problems. Finally, all of this work may be for naught if the product or idea is not effectively tested against reality. It is particularly important to use reality tests because the human mind does not function very logically and is prone to self-delusion. For this reason, internal reality checks are useful. We may check for the coherence of an idea or theory, examine the premises behind an idea or product, and use if–then conditional reasoning to verify the implications of our idea or product. It is also the case that various types of reasoning are suspect, such as arguments by opposition (exclusionary logic), reasoning from inadequate data, faulty generalization, and deductions that do not follow from the data. These flaws in reasoning should be checked for and corrected if they affect a product or idea. It is also necessary to test a product or idea on its merits. The first thing an inventor or creator should do is try to break the creative product, either physically or by finding conditions under which it is not valid or fails. It is also imperative to test a creation against some type of external data or standard. Techniques such as experiments, prototypes, scale models, peer review, opinion polls, market testing, etc. can be crucial in this regard.

Style and attitude affect performance at all four of the above stages. It was shown that different styles of attacking problems prove more successful for certain types of problems than for others. For example, we may contrast the tinkerer with the visionary. The tinkerer prefers to improve an existing product or process whereas the visionary prefers to start from scratch and create something new and better. It is critical that a person with a defined style seek out work environments or problem domains that are amenable to their style of working or else they will not be very successful no matter how hard they work and no matter how smart they are. A visionary in an entry level job is not likely to be given the latitude to apply his or her vision in practice. Style can interact with the different stages of problem solving and affect productivity in a direct way. For example, the visionary may neglect elaboration and reality testing and leap right into declaring that an idea is a finished product. In the software realm, this is called vaporware. On the other hand, the tinkerer may fiddle forever with the details and never ship out a completed product. Attitude also interacts with these four stages. For example, we saw that the person who lacks a little detachment may be so afraid of failure that they never dare to look into a wild idea that they have. The person who lacks humility may ignore all feedback from reality checks that might help them correct an erroneous idea. Most pernicious of all, a person who is dishonest may ignore hints that the product is dangerous or the idea flawed. Thus aspects of personal style and attitude can aid the person to be more effective or lead them astray if there is a mismatch between style and task or if their attitude is flawed.

It is useful to examine the question of what tasks might provide useful training in strategic thinking. Tasks that promote intense concentration, such as chess, promote the intensity necessary to carry out long chains of reasoning. Tasks that teach hierarchical organizational skills, such as computer programming, can help one learn to master complex sets of information. Familiarity with maps, charts, diagrams, and blueprints also provides useful mental tools that can be applied in other domains. Familiarity with odds and chance is also quite useful. Surprisingly, skill in team sports is not particularly helpful, because in real life the rules change too much and the opponent is not identified by a different color jersey. Finally, the practice of building things can provide invaluable experience for strategic problem solving.

Based on these ideas, it is possible to be a more effective problem solver. There are three components of this: problem finding, the use of specific tools of thought, and generating solutions.

Problem finding utilizes the fact that problems worth solving usually generate certain symptoms, including paradox and contradiction, or may be found by considering changes in perspective and scale. Paradox is a particular configuration of facts that points to an opportunity. A paradox means either that the general understanding of a phenomenon is faulty, or that our picture is incomplete. In either case, if one can recognize a paradox that others miss then there is an opportunity for discovery, for innovation, or for profit. Bottlenecks in systems can cause paradoxes, and represent possible control points for a system.

Consideration of perspective and scale provides a number of insights. Many times the nature of a problem can only be discerned with the proper perspective, which may simultaneously provide a solution. Perspective also provides insights into the causes of misunderstandings that arise in human systems. Changes in perspective (e.g. of a corporation) will show different aspects of functioning, and may show a system to be performing well in some ways and poorly in other ways. We may recognize viewpoint and time perspectives. Scale problems arise in many contexts, from growing organizations to the scaling up of architectural models or production processes. Changes of scale occur with changes in size, speed, and complexity, and general laws of scaling exist.

There are a number of general tools of thought that are generally useful, including evolutionary refinement, analogy, classification, the detective model, failure analysis, and models of cycles, spirals, nets, and webs. Evolutionary refinement is directly analogous to the evolutionary process. It begins with a functional, but crude, solution, and then successively refines it. Software prototyping is a good example. Analogy is a very powerful problem solving tool, though it is not the whole picture, as some maintain. Analogy helps us locate in memory similar past problems to the one at hand, which may help generate a solution. It can also provide models, comparisons, and bridges across disciplinary gulfs. The detective model is very useful for narrowing down lists of candidates (causes of a disease, suitable homes to purchase, etc.) based on sets of criteria. Classification is an essential problem solving tool. Classifications organize information, models, or theories so that they can be applied more effectively. Problem solving in many cases involves a reclassification into new categories. Decomposition into component parts is an important type of classification. The generation of scenarios is a useful type of problem solving strategy. It allows plans to be constructed so that their feasibility may be examined. The What if? type of scenario is useful for testing out the implications of some action that has never previously occurred. This represents the tracing out of the consequences of a change in the normal rules. In failure analysis, we examine what could possibly go wrong, and try to design against such failures. Failure analysis is common in building design and in warfare, but can profitably be applied in other areas as well. We may characterize systems that fail catastrophically as brittle, and those that fail gracefully as robust. When human lives are at stake, it is best to avoid brittle designs. It is argued that many processes may be modeled with cycles or spirals (e.g. seasonal cycles of the economy linked to agriculture). Utilizing these models may allow such systems to be better understood. A transformation occurs when something moves from one state to a qualitatively different one (e.g. by marriage or incorporation). Recognition of transformations and the acts by which they are brought about allows them be understood and manipulated to advantage. Finally, nets and webs are useful constructs for depicting complex relationships.

Generating a solution involves getting a grip on the problem, like the small boy leading the water buffalo by a ring in its nose. It is not sufficient to know lots of facts about something. New knowledge is only generated or systems controlled by those who apply leverage to control points. Knowing where to push and how is the key. An understanding of constraints, complexity, feedbacks, information, and bottlenecks provides this knowledge.

Constraints and tradeoffs prove to be present in most design, engineering, and systems problems. They provide insight into the structure of a system and provide avenues of attack for understanding or controlling the system. A major focus of industrial activity is the overcoming of constraints with new designs, materials, and processes. New lightweight materials, for example, have improved the gas mileage of cars. New production processes and new chip designs continually improve the performance of computers. Constraints and tradeoffs can be recognized as critical to the design of biological systems. Considerable progress has been made in biology by examining the tradeoffs resulting from physical and metabolic constraints. We can clearly recognize constraints in economic systems. For example, there is the well-known tradeoff between risk and yield in the stock market. The constraint that enforcing rules must have a cost is often overlooked. Attempts to circumvent tradeoffs often lead to deficit financing. Finally, we may recognize constraints in the context of personal living. It is difficult, for example, to be in two places at once. Information and new technologies can help alleviate constraints in these varied domains, but constraints will always exist and acting as they do not will always lead to trouble.

Complexity is an important attribute of many problems. Single causes are rarely at the root of phenomena. Various tools can help tease out complex relationships. Tracing the threads of causation is useful, particularly for distributed causation. We may trace out causation by noting that nothing is free (which helps us account for all costs), by observing that everything goes someplace (which focuses attention on ultimate destinations or outcomes), and by realizing that entropy is always involved (which draws attention to disorder produced by a process). Complexity per se is an important attribute of many systems. Complex economies appear to be more stable than simple ones because there are more industrial sectors which are less likely to be synchronized in their cycles and because there are mechanisms such as savings accounts and lines of credit that buffer against adverse periods.

Feedback exists in any dynamic system and can lead to stable or unstable behavior. An understanding of positive and negative feedback and of lags can provide insights into the dynamics and control points in a system. Information is a specific type of feedback that is particularly important for improving professional performance and for managing organizations. Errors of communication and of converting data into information lie at the root of many problems. Merely correcting how information is processed can often lead to improved system or personal performance.

Bottlenecks represent very useful control points. In a business setting, a bottleneck in production represents an opportunity to improve productivity. It must be cautioned that removal of one bottleneck will certainly move the bottleneck elsewhere, since something is always limiting to production, but the new bottleneck may be distributed and not necessarily easy to find. To exercise control, it is often necessary to create a bottleneck, such as a central point for financial control or for clearance of public statements by an organization. High throughput systems are particularly prone to bottlenecks. It is useful to eliminate bottlenecks that are not actual control points, such as holdups in procurement of supplies that do not actually serve a control purpose (it serves no purpose to hold up operations because paper is on back order).

The strategic thinking process thus represents a synthesis of knowledge on how to generate novel ideas more effectively, how to convert ideas into products, how to utilize conceptual tools, and of how complex systems function. This synthesis puts the solution of complex problems within reach and can increase the leverage that the individual problem solver can exert.

Science is not a totally solitary endeavor nor is it just mental. Active manipulation of the world is necessary. Scientists must undertake an active research program, interact with others, and communicate their results. All of these activities are nontrivial and offer the potential for self-deception, abuse of power, bias, and other problems. Thus the final section of the book focused on the social dimension of science in practice.

The overall view expressed in this book is that scientists are largely uncoached and are rarely introspective. They spend a lot of time studying their disciplinary subject matter, but almost no time learning strategies of problem solving. The work of scientists could be enhanced, I believe, by application of the ideas presented in this book.

References

Adler, M. J. 1985. *Ten Philosophical Mistakes*. New York: Collier Books.

Albert, R. S. 1975. Toward a behavioral definition of genius. *American Psychologist* **30**: 140–151.

Anderson, C. 1992. Writer's cramp. *Nature* **355**: 101.

Arieti, S. 1976. *Creativity: The Magic Synthesis*. New York: Basic Books.

Arnott, R. and K. Small. 1994. The economics of traffic congestion. *American Scientist* **82**: 446–455.

Bailey, J. 1995. The recycling myth. *The Wall Street Journal*, Jan. 19.

Bauer, H. H. 1992. *Scientific Literacy and the Myth of the Scientific Method*. Urbana, IL: University of Illinois Press.

Biondi, A. M. 1980. About the small cage habit. *Journal of Creative Behavior* **14**: 75–76.

Boorstin, D. J. 1983. *The Discoverers*. New York: Random House.

Boorstin, D. J. 1992. *The Creators*. New York: Random House.

Boxenbaum, H., F. Pivinski, and S. J. Ruberg. 1987. Publication rates of pharmaceutical scientists: application of the Waring distribution. *Drug Metabolism Reviews* **18**: 553–571.

Bylinsky, G. 1994. The digital factory. *Fortune*, Nov. 14, pp. 92–110.

Collins, H. and T. Pinch. 1993. *The Golem*. Cambridge, UK: Cambridge University Press.

Crease, R. P. 1992. The trajectory of techniques: lessons from the past. *Science* **257**: 350–353.

Darwin, C. 1881. *The Formation of Vegetable Mould, Through the Action of Worms, With Observations on Their Habits*. London: John Murray.

Darwin, F. (ed.) 1958. *The Autobiography of Charles Darwin and Selected Letters*. (Original work published 1892.) New York: Dover.

DeCicco, J. and M. Ross. 1994. Improving automotive efficiency. *Scientific American*, Dec., pp. 52–57.

Drucker, P. F. 1985. *Innovation and Entrepreneurship*. New York: Harper & Row.

Feynman, R. P. 1984. *Surely You're Joking Mr. Feynman*. New York: Norton.

Friend, T. 1995. That operation you're getting may be experimental. *USA Today*, Sept. 13.

Galvin, R. 1995. *Alternative Futures for the Department of Energy National Laboratories* (Galvin Report). Washington, DC: U.S. Department of Energy.

Gardner, H. 1983. *Frames of Mind: The Theory of Multiple Intelligences*. New York: Basic Books.

Gibbs, W. W. 1994. Software's chronic crisis. *Scientific American*, Sept., pp. 86–95.

Giere, R. N. 1994. The cognitive structure of scientific theories. *Philosophy of Science* **61**: 276–296.

Gilbert, T. F. 1978. *Human Competence.* New York: McGraw-Hill.

Gleick, J. 1987a. *Chaos: Making A New Science.* New York: Penguin Books.

Gleick, J. 1987b. New images of chaos that are stirring a science revolution. *Smithsonian,* Dec., pp. 122–135.

Hadamard, J. 1949. *The Psychology of Invention in the Mathematical Field.* Princeton, NJ: Princeton University Press.

Hall, S. S. 1992. How technique is changing science. *Science* **257**: 344–349.

Hunter, M. S. 1978. Exiting creative researchers from graduate education. *Journal of Creative Behavior* **12**: 209–213.

Jamison, K. R. 1995. Manic-depressive illness and creativity. *Scientific American,* Feb., pp. 62–67.

Johnson, P. 1988. *Intellectuals.* New York: Harper & Row.

Jones, G. and K. Douglas. 1994. The quiet genius who decoded life. *New Scientist,* Oct. 8, pp. 33–35.

Kellow, A. 2007. *Science and Public Policy.* Northampton, MA: Edward Elgar.

Koestler, A. 1964. *The Act of Creation.* New York: Macmillan.

Kuhn, T. S. 1970. *The Structure of Scientific Revolutions.* Chicago, IL: University of Chicago Press.

Lackey, R. T. 2007. Science, scientists and policy advocacy. *Conservation Biology* **21**: 12–17.

Langley, P. and R. Jones. 1988. A computational model of scientific insight. In R. J. Sternberg (ed.), *The Nature of Creativity.* Cambridge, UK: Cambridge University Press, pp. 177–201.

LeVay, S. 2008. *When Science Goes Wrong.* New York: Penguin Group.

Lightman, A. and O. Gingerich. 1992. When do anomalies begin? *Science* **255**: 690–695.

Loehle, C. 1987. Hypothesis testing in ecology: psychological aspects and the importance of theory maturation. *Quarterly Review of Biology* **62**: 397–409.

Loehle, C. 1988a. Philosophical tools: potential applications to ecology. *Oikos* **51**: 97–104.

Loehle, C. 1988b. Tree life history theory: the role of defenses. *Canadian Journal of Forest Research* **18**: 209–222.

Loehle, C. 1989. Catastrophe theory in ecology: a critical review and an example of the butterfly catastrophe. *Ecological Modelling* **49**:125–152.

Loehle, C. 1990. A guide to increased creativity in research: inspiration or perspiration? *BioScience* **40**: 123–129.

Loehle, C. 1994. *On the Shoulders of Giants.* Kidlington, Oxfordshire, UK: George Ronald.

Loehle, C. 1996. *Thinking Strategically.* Cambridge, UK: Cambridge University Press.

Loehle, C. 2006. Control theory and the management of ecosystems. *Journal of Applied Ecology* **43**: 957–966.

Loehle, C. and G. Wein. 1994. Landscape habitat diversity: a multiscale information theory approach. *Ecological Modelling* **73**: 311–329.

Mann, C. 1990. Meta-analysis in the breach. *Science* **249**: 476–480.

Margolis, H. 1987. *Patterns, Thinking, and Cognition.* Chicago, IL: University of Chicago Press.

McDonald, K. A. 1990. Researchers increasingly worried about unreliability of big-science projects. *Chronicle of Higher Education* **36**:1, A8–A9.

Medawar, P. B. 1967. *The Art of the Soluble.* Oxford, UK: Oxford University Press.

Parker, G. 1989. Predicting the productive research psychiatrist. *British Journal of Psychiatry* **154**: 109–112.

Peters, T. and N. Austin. 1985. *A Passion For Excellence*. New York: Warner Books.

Peters, T. and R. H. Waterman, Jr. 1982. *In Search of Excellence*. New York: HarperCollins.

Petroski, H. 1992. *The Evolution of Useful Things*. New York: Vintage Books.

Pfennig, D. W. and P. W. Sherman. 1995. Kin recognition. *Scientific American*, June, pp. 98–103.

Popper, K. R. 1963. *Conjectures and Refutations: The Growth of Scientific Knowledge*. New York: Harper & Row.

Richman, L. S. 1993. Why the economic data mislead us. *Fortune*, March 8, pp. 108–114.

Root-Bernstein, R. S. 1989. *Discovering*. Cambridge, MA: Harvard University Press.

Sidorowich, J. J. 1992. Repellors attract attention. *Nature* **355**: 584–585.

Simon, J. L. 1980. Resources, population, environment: an oversupply of false bad news. *Science* **208**:1431–1437.

Simonton, D. K. 1988. *Scientific Genius*. New York: Cambridge University Press.

Skinner, B. F. 1959. A case study in scientific method. In S. Koch (ed.), *Psychology: A Study of a Science*. New York: McGraw-Hill, pp. 359–379.

Stanovich, K. E. 1992. *How to Think Straight About Psychology*. New York: HarperCollins.

Sternberg, R. J. 1988. *The Nature of Creativity*. Cambridge, UK: Cambridge University Press.

Thelan, H. A. 1972. *Education and the Human Guest*. Chicago, IL: University of Chicago Press.

Treisman, M. 1995. The multiregional and single origin hypotheses of the evolution of modern man – a reconciliation. *Journal of Theoretical Biology* **173**: 23–29.

Tsonis, A. A. and J. B. Elsner. 1992. Nonlinear prediction as a way of distinguishing chaos from random fractal sequences. *Nature* **358**: 217–220.

Waldrop, M. M. 1992. *Complexity*. New York: Simon & Schuster.

Watson, J. D. 1968. *The Double Helix*. New York: New American Library.

Whimbey, A. and L. S. Whimbey. 1976. *Intelligence Can Be Taught*. New York: Bantam.

Index

Made in the USA
Monee, IL
09 July 2021